Embedded Artificial Intelligence

Bin Li

Embedded Artificial Intelligence

Principles, Platforms and Practices

Bin Li
CTO Office
Beijing DataRealm Technology
Beijing, China

ISBN 978-981-97-5037-5   ISBN 978-981-97-5038-2   (eBook)
https://doi.org/10.1007/978-981-97-5038-2

Jointly published with Tsinghua University Press

This Springer imprint is published by the registered company Springer Nature Singapore Pte Ltd.
The registered company address is: 152 Beach Road, #21-01/04 Gateway East, Singapore 189721,
Singapore

If disposing of this product, please recycle the paper.

Preface

Embedded artificial intelligence is a big topic!

Let's reflect on the origins of the term "artificial intelligence": "The science and engineering of creating intelligent machines, especially intelligent computer programs"—John McCarthy, 1956. Essentially, from the outset of artificial intelligence, the aim has been to create intelligent machines. This goal carries two implications: firstly, designing machines specifically for intelligence, and secondly, integrating intelligence into existing machines. While our efforts have predominantly focused on the former, the aspiration to embed intelligence into machines has been present from the beginning.

Typically, when discussing artificial intelligence, thoughts often turn to robots. In the extravagant realms of science fiction, AI embodies mechanical suits transforming ordinary individuals into heroes (like Iron Man), mutated entities threatening global destruction (like I, Robot), or endearing mechanical companions (like The Robotic Butler). The desire for machines resembling humans, capable of hearing, seeing, speaking, acting, and even contemplating, has persisted. Yet, achieving such machines remains profoundly challenging. With the emergence of ChatGPT, there's a dawn of hope—computers can now speak like humans! However, a significant challenge remains: can we achieve such marvels on machines like humanoid robots, with dimensions and power consumption comparable to humans? ChatGPT's training requires over 30,000 GPUs, with a total power exceeding 10 million watts and a daily electricity cost surpassing $50,000. In contrast, the human brain occupies merely about 1.5 L and operates on less than 20 W. If we aim to embed artificial general intelligence into robots, cars, or even smaller devices like drones, phones, smart appliances, or IoT devices, we must overcome substantial hurdles. This is precisely the topic explored in this book *Embedded Artificial Intelligence*.

The initial concept for this book emerged in 2018, inspired by a clever child's idea to create an autonomous flying sun umbrella to provide shade when he is playing. It was a cool idea but incredibly challenging! At that time, academic research regarding embedded artificial intelligence was just beginning, with scattered achievements yet to undergo large-scale practical validation. Implementing such advanced functionality in a tiny processor for drones was still unrealistic. Nonetheless, we believe that one day it will be realized, so let's strive for that dream.

Over the following years, we delved into various possibilities to realize this dream. Thankfully, it's not just our personal aspiration; it's a collective dream of the entire industry. Research on embedded artificial intelligence surged like mushrooms after rain, with lightweight algorithms emerging, significant progress in model compression techniques, and AI acceleration chips tailored for embedded systems hitting the market. Thus, as we explored, summarized, and wrote, we roughly completed the first part of this book: the principles. Sharing it with some friends garnered a bit of encouragement, prompting the start of the second and third parts: the platforms and practices. Both parts compile data gathered during our efforts to realize the autonomous flying sun umbrella. Though they may be somewhat outdated by the time this book hits the market, they are comprehensive enough to offer valuable insights to readers.

In 2023, the Chinese version of this book was published. We were delighted to observe the rapid development of various platforms for embedded artificial intelligence and the blossoming of embedded AI practices. Discussions on embodied intelligence have begun, envisioning a brighter future with the new generation of autonomous vehicles and robots. Underlying these imaginations is the rapid advancement of embedded artificial intelligence. Key technologies in embedded AI, such as model compression, not only shine in the embedded domain but also serve as a crucial factor in deploying large-scale models like ChatGPT at a low cost. For models with trillions of parameters, even the largest GPUs become small chips! In the future, people will undoubtedly seek to implement brain-scale models on machines with the size and energy consumption of the human brain, presenting an entirely new challenge. The future of embedded artificial intelligence knows no bounds!

Finally, I would like to express gratitude to the mentors and friends who guided and assisted in the making of this book. This book comprehensively synthesizes previous achievements in the field of embedded artificial intelligence. Thus, first and foremost, thanks are due to the scholars who conducted pioneering research in this field, including but not limited to Xipeng Shen, Song Han, Shaoshan Liu, Mingxing Tan, Menglong Zhu, Forrest Iandola, François Chollet, Andrew Howard, Bert Moons, Daniel Bankman, Marian Verhelst, and others. This list is bound to omit some contributors due to the rapid pace of developments in this field; your understanding is appreciated. Secondly, I extend thanks to friends and colleagues who provided feedback and suggestions during the writing process. It's your tireless encouragement and assistance that enabled me to persist in completing this extensive book. Lastly, I must thank my family, my wife, Lucia, who meticulously translated every word of this book, and our child, Jerome, who contributed inspiration to this writing!

Beijing, China   Bin Li

Contents

Part I
Principles

Chapter 1
Embedded Artificial Intelligence

Abstract What is embedded artificial intelligence? Why do you need embedded artificial intelligence? How to implement embedded artificial intelligence? What are the challenges of implementing embedded artificial intelligence? With these questions, we defined the topics to be studied in this book. After comparing the two implementation modes of embedded artificial intelligence: cloud computing mode and local mode, we clarified the necessity and technical challenges of implementing the local mode and outlined the five essential components needed to overcome these challenges and achieve true embedded AI.

Keyword Embedded AI

1.1 What Is Embedded Artificial Intelligence?

When thinking of artificial intelligence, most of us probably think of robots and computers. In the magnificent imagination of science fiction novels and movies, AI is a supercomputer that creates everything (such as The Matrix), a robot that destroys the world (such as Mechanical Enemy) or a spaceship that cruises in the vast universe (such as 2001: A Space Odyssey). However, it is not. In the early days, artificial intelligence was more of an intelligent system with decision-making capabilities. For example, there were applications designed specifically for spelling and grammar checking. When these applications were first introduced for computers, they were considered highly intelligent. These applications were among the earliest forms of AI, and today they are so commonplace that they no longer carry the title of AI. However, although AI is now more diverse in appearance, most of the time it is still just a large, complex computer software that implements some kind of "intelligent" algorithm to solve problems. For example, AlphaGo defeats humans, or AI characters in games, or translation software that understands all languages. But as we dream, AI has begun to be combined with devices to build truly intelligent machines. This is exactly the subject of this book, embedded artificial intelligence (abbreviated as embedded AI).

© Tsinghua University Press 2024

B. Li, *Embedded Artificial Intelligence*,

https://doi.org/10.1007/978-981-97-5038-2_1

So, what is embedded artificial intelligence?

Let's review the origin of the name artificial intelligence,

The science and engineering of making intelligent machines, especially intelligent computer programs—John McCarthy, 1956.

In other words, from the early days of artificial intelligence, we have wanted to create intelligent machines. This sentence has two meanings. First, create machines specifically designed to achieve intelligence. Second, create machines with embedded intelligence. In the past few years, we have been working hard toward the first goal, but in fact from the beginning we have been looking forward to "embedding" intelligence into devices!

Traditionally, an embedded device refers to a device that embeds a computer system to achieve specialized functions and real-time computing performance. Compared with a general-purpose computer, on the one hand, it is "smaller." It is not universal and only meets certain specific needs. Therefore, it uses a processor with weaker performance, such as a microcontroller, and the memory is also limited by the size of the device. And smaller, on this basis, the software system running on the device is also lightweight, generally using a compact embedded operating system, and the application software only completes specific and limited functions and is smaller in scale. Benefiting from this streamlined software and hardware, embedded devices also consume less power, and some can be powered by batteries. But on the other hand, it is "bigger." Many embedded systems are mechatronics devices. In addition to computer systems, they also have sensors, execution components, etc. This allows it to interact more with the external world and achieve connection with the physical world. Another characteristic of embedded systems is real-time nature. They must respond to inputs within a limited time, just like a real living organism.

With the advancement of computer software and hardware technology, these embedded devices have become more and more powerful and "intelligent." With the advancement of computer software and hardware technology, these embedded devices have become more and more powerful, and their "intelligence" is getting higher and higher, such as smart phones, which have made significant progress in recent years, and their performance has caught up with general-purpose computers and can complete most of the daily computing tasks. Another example is smart home hardware, which can proactively turn on the air conditioner before the owner returns home, achieving the ability of independent analysis and decision-making to a certain extent. So, can it be said that these "smart hardware" has implemented embedded artificial intelligence?

Strictly speaking, not yet.

The artificial intelligence introduced earlier in this book is implemented on a general-purpose computer. In particular, this artificial intelligence is implemented by some intelligent software, which may be a chess program, an expert system, a deep neural network, or a robot operating system，running on a powerful general-purpose computer, such as GPU or supercomputer. Of course, this consumes a lot of power. Then, the goal of embedded artificial intelligence in the strict sense is to achieve equivalent artificial intelligence on the limited, special hardware resources and strict power budget of embedded devices. Note that "equivalent" here mainly refers to the equivalence of functions. For example, embedded devices can implement image recognition, speech recognition and other functions like general-purpose computers, but their performance can be slightly compromised.

Broadly speaking, we refer to the ability of autonomous analysis and decision-making implemented on embedded devices as embedded artificial intelligence. In a narrow sense, **embedded artificial intelligence is artificial intelligence implemented on embedded devices that is equivalent to artificial intelligence implemented on general-purpose computers.**

In recent years, the concept of **embodied intelligence** has been proposed, which refers to intelligent agents that have a body and support physical interaction, such as home service robots, autonomous vehicles, etc. In contrast, Disembodied AI refers to artificial intelligence that does not have a physical body and can only passively accept data collected and produced by humans. Embodied intelligence is closely related to embedded artificial intelligence. It can be considered as a specific implementation of embedded artificial intelligence. It has more clear and specific requirements for embedded devices, that is, having a body. In this sense, embodied intelligence is a subset of embedded intelligence.

The relationship between artificial intelligence, embedded artificial intelligence, and embodied intelligence is shown in Fig. 1.1.

How to implement artificial intelligence on embedded devices? The main implementation method of embedded artificial intelligence studied in this book is still neural network, especially deep neural network. When they are implemented on embedded devices, we call them embedded neural networks or embedded deep neural networks. In the later chapters of this book, there is no strict distinction between these two terms, and they both refer to embedded deep neural networks.

Fig. 1.1 The relationship between artificial intelligence, embedded artificial intelligence, and embodied intelligence

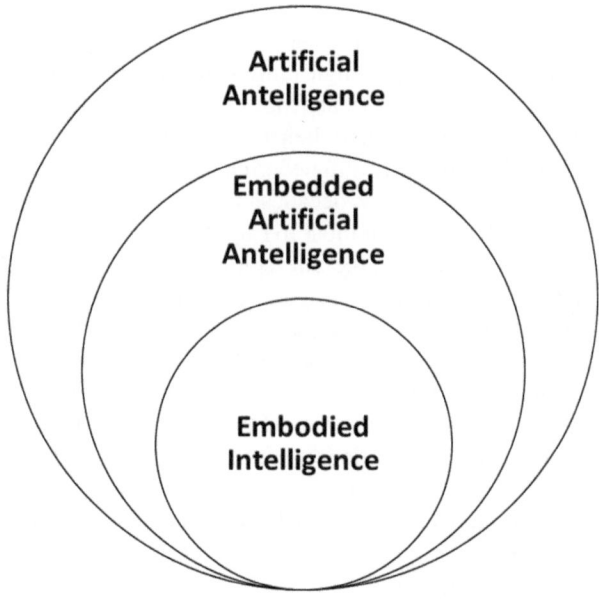

So, what is the difference between neural networks implemented in embedded devices and neural networks implemented on general-purpose computers? Can general-purpose neural networks be deployed directly into embedded devices? Are embedded neural networks just scaled-down versions of general-purpose neural networks? Can embedded neural networks achieve artificial intelligence equivalent to that achieved on general-purpose computers?

We will take these questions into the later chapters of this book. Whatever the answer to the above question, one thing is for sure: Embedded AI is needed.

1.2 Why Do You Need Embedded Artificial Intelligence?

If a general-purpose computer is like a brain, then an embedded computing device is like a complete living body with a brain, sense organs, and limbs. In the past, the brains of embedded devices were not very developed and could only complete some programmed tasks set by humans in advance. But if we give artificial intelligence to machines, let them "live," and let their "minds" independently perceive the environment from the "sense organs, "and command the "limbs" to adapt to the environment and transform the environment, then isn't such a machine the intelligent machine that we have envisioned since the beginning of artificial intelligence that can hear, see, speak, do, and think? Only then can we say that we have achieved true artificial intelligence!

It is this dream that has led scientists and engineers around the world to devote themselves to research and development in the field of embedded artificial intelligence.

From another perspective, embedded devices are everywhere, and they mostly use SoC chips (system-on-a-chip), which integrate microprocessors, memory, I/O interfaces, etc. into a single chip, which is the heart of embedded devices. Their low energy consumption means they can run on batteries for months and require no heat sink, and their simplicity helps reduce the overall cost of the system. More than 10 billion SoC chips are shipped globally every year.

Over the past few decades, the computing power of SoCs has continued to increase. However, in most IoT applications, they simply send data from sensors to the cloud. Therefore, the SoC is idle most of the time. On the other hand, SoCs have made rapid progress this year, and many SoCs have integrated dedicated neural network computing processors, such as NPUs. These SoCs are not just sensors and communicators but can also perform local neural network computations.

Imagine how much computing power is wasted in the real world with tens of billions of such devices deployed in the real world! If we can harness this power and empower tens of billions of edge devices with true intelligence, our world will become a true AI world.

If the above dream can be realized, embedded artificial intelligence will be everywhere. Specialized intelligent machines developed for specific tasks will continue to emerge. Some are simple (such as smart switches), some are complex (such as autonomous vehicles); some perform a single function (such as license plate recognition cameras), and some can perform multiple functions (such as smart-phones); some have only "sense organs" (such as IoT sensors), and some with all sense organs and limbs (such as robots). These intelligent machines will become our assistants at work and partners in life.

Over the past few years, embedded systems have reached a certain level of " intelligence ." More and more smart devices are being released every day and every month. Artificial intelligence on embedded systems has begun to move from very basic forms to complexity and polymorphism, just like the process of biological evolution.

For example, smart home lighting systems automatically turn on and off based on whether someone is in the room. On the surface, this system isn't a big deal. But when you think about it more deeply, you realize that this system is making deci-sions on its own. Based on the input from the sensor, the SoC decides whether to turn on the light or not. Isn't this a very basic form of AI in embedded systems?

There are also many examples of simple forms of AI in embedded systems. But what about the future? Are we about to have embedded systems with AI that can completely replace human jobs?

Let's look at how AI and embedded systems work together and how they evolve.

1.2.1   Image Identification

Soon, the convergence of artificial intelligence and embedded systems will lead to huge advances in image and video recognition. Advances in embedded technology will help us build imaging devices with higher processing power and smaller foot-prints. At the same time, AI will provide the much-needed algorithms needed for real-time image and video recognition. The implementation of these smart imaging devices for public safety will be beneficial as it will detect potentially dangerous behavior. Such systems will also be adopted to improve inventory management in factories, monitoring of transportation systems, and the development of industrial automation. For example, license plate recognition cameras have been widely deployed in parking lots. These cameras have embedded image recognition algo-rithms that can quickly and accurately obtain license plate numbers to complete access control and billing.

1.2.2   Self-Driving

Embedded systems and cars are much closer than you think. Navigation systems, airbag deployment mechanisms, anti-lock braking systems, and many more are based on embedded systems. But bringing AI to cars will be a real game-changer. Self-driving cars have been under development for the past few years and are also undergoing numerous field trials. Tech giants like Google, Tesla, and Uber are investing billions in research and development with an eye on creating a driverless future. We won't be giving up driving anytime soon though, it could be 10–15 years before you see robot cars cruising the streets. In the process, AI will gradually be introduced into traditional cars, adding more and more autonomous functions. For example, some cars have implemented automatic parking functions to help less skilled drivers park the car into a parking space. Soon, advances in embedded systems will help manufacturers put powerful sensors on boards. This will enable cars to automatically deploy countermeasures, such as automatically braking in emergency situations to avoid traffic accidents.

1.2.3   Dangerous Work

Some of the most dangerous jobs in factories are already taken care of by machines. Thanks to advances in embedded electronics and industrial automation, we have powerful microcontrollers running entire assembly lines in manufacturing plants. However, most of these machines are not fully automated and still require some kind of human intervention. However, now is the time to introduce AI, which can help engineers design truly intelligent machines that can operate with zero human intervention. One such area is the development of bomb-disposal robots. AI-equipped machines can take over tasks such as manufacturing, drilling, and welding of potentially hazardous chemicals.

1.2.4   Internet of Things

The introduction of artificial intelligence will also greatly benefit the Internet of things. We will have smart automation solutions to save energy, improve cost efficiency, and eliminate human error. According to Gartner, more than 20 billion IoT devices will be in use by 2020, and these devices will generate more than 500 ZB of data every year. With more and more technological advancements, this number is expected to continue to grow dramatically. To process the massive data generated on these massive devices is beyond the reach of humans, and artificial intelligence is undoubtedly needed to meet this challenge. In the past, these IoT devices were just embedded devices with built-in sensors and simple controllers to complete some programmed tasks, such as smart light poles that determine their own switches by sensing the intensity of ambient light. In the future, these smart light poles may have "eyes" that dim the light to save energy when no one is passing by at night and illuminate their path when someone is about to pass by. Furthermore, these smart light poles may have the function of human–machine dialogue, introducing information about surrounding shopping malls and restaurants to passers-by, becoming a ubiquitous guide in the city.

1.2.5 Smart Phone

Smartphones, as personal intelligent assistants, have been integrated with artificial intelligence a long time ago. Before the emergence of touch-screen mobile phones, mobile phones that supported voice commands and handwriting recognition had been developed to achieve a more friendly human–computer interaction interface. However, at that time, the traditional pattern recognition algorithm is used, but its recognition rate is not enough to meet the demanding needs of users. Today, smartphones have almost become a new organ of the human body. They carry millions of APPs and realize a variety of functions. The introduction of deep neural network algorithms will achieve better image, voice, text recognition, and other capabilities. This enables more intelligent applications. These scenarios include object recognition, gesture recognition, motion detection, sentiment analysis, natural semantic recognition, music tagging, and more. For example, how to trigger a mobile phone to take a selfie is a problem that has not been completely solved. Touching the screen, selfie stick, Bluetooth remote control, voice commands, etc. will introduce additional actions, making the look less natural. But if the mobile phone can recognize people's expressions, gestures, and body postures, it can trigger the mobile phone at the most appropriate time and capture the most touching moments like a professional videographer.

In addition to these areas, the fusion of artificial intelligence and embedded systems will bring many other opportunities. Such as medical care, logistics, fire protection, agriculture, communications, military, etc.

In the final analysis, the fusion of artificial intelligence and embedded systems will give intelligence to all things, and will allow machines to replace us in completing time-consuming, laborious and even dangerous tasks, thus improving our lives, improving the efficiency of our work, and engaging in tasks that humans are incapable of. Even go to alien planets to open human living space. Embedded artificial intelligence will change the future of mankind!

1.3 Initial Attempt: Cloud Computing Mode

Artificial intelligence has traditionally relied on high-performance computing power provided by server clusters for large-scale, data-intensive model training and inference in the cloud. Due to the significant increase in GPU hardware performance, artificial intelligence, especially deep neural networks, are being applied to more and more business applications, including finance, education, medicine, security, etc. However, the problem with such algorithms is that they are greedy consumers of data and like to complicate the problem. Only with larger data sets and more intensive computing power can more accurate and useful results be obtained.

Therefore, until recently, deep neural networks relied on big data and cloud computing and had to run on energy-intensive servers or even supercomputers with AI accelerators (such as GPU /TPU). This computer hardware is bulky, energy-hungry, difficult to move, and expensive. However, the current applications of deep neural networks mostly focus on the fields of computer vision and hearing. The problems it solves, such as face recognition, license plate recognition, natural language translation, voice control, etc., often require lightweight, green, energy-saving, and easy to move, low-cost embedded computing devices (including mobile computing devices such as mobile phones) to complete. However, the computing power of these embedded devices is several orders of magnitude lower than that of GPU/TPU, and the memory is also limited. It is obviously not enough to run deep neural networks. So, how to resolve this contradiction?

Many people will naturally think of using cloud computing! Wouldn't it be wonderful to stream data, such as pictures, video streams, and audio to the cloud, and let the powerful cloud computing center complete deep neural network tasks? In the beginning, people did do this. For example, in smart home scenarios, smart hardware only served as sensors and controllers to collect data from the field. The real intelligence was completed by the cloud computing center and issued the results of AI operations (instructions) to control the intelligent hardware to complete the task. This model is the first stage of embedded artificial intelligence and can be called **cloud computing mode.** As shown in Fig. 1.2.

In this mode, the embedded device itself only completes simple data collection, communication, command execution, etc. AI hardware (GPU/TPU), AI algorithms, and AI applications are all deployed in the cloud, and the embedded devices invoke their capabilities through remote interfaces to achieve advanced intelligence.

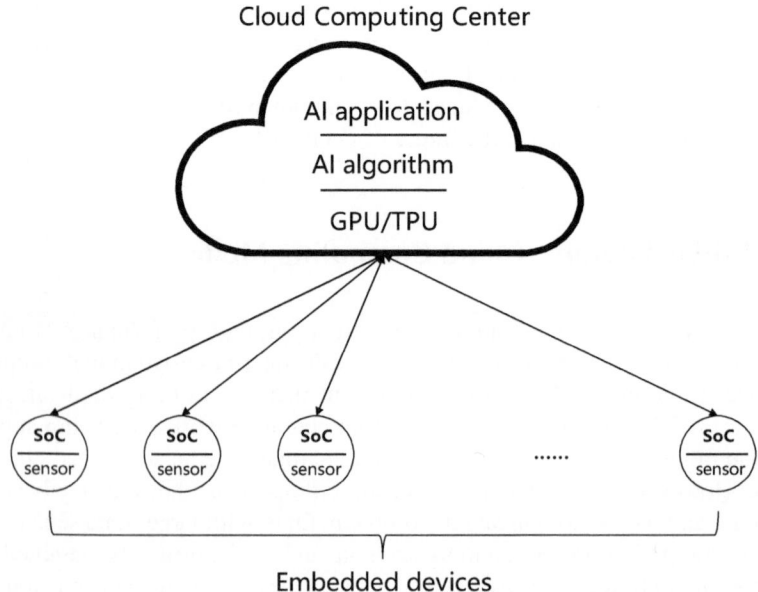

Fig. 1.2 Cloud computing mode of embedded artificial intelligence

This model has obvious advantages:

1. It allows numerous embedded devices to share professional and expensive artificial intelligence hardware (such as GPU/TPU), reducing the cost of a single embedded device.
2. With the help of mature artificial intelligence technology in the cloud, embedded artificial intelligence applications can be quickly developed and deployed.
3. Embedded AI applications are easy to upgrade and maintain because the AI programs are deployed in the cloud.
4. Elastic computing can be achieved. When the current hardware cannot meet the performance requirements, more resources can be applied for.
5. Users can purchase and use artificial intelligence cloud services on demand.

Unfortunately, this model doesn't work in many cases.

Let's start with an accident involving Amazon's Alexa voice assistant.

Alexa is the AI voice assistant on the Echo smart speaker, which is a smart speaker sold by Amazon. The appearance of Echo is no different from ordinary Bluetooth speakers, and it does not have any screen. The only interaction method is voice. Through the Alexa voice assistant, users can play music, query information, and even control various smart home devices through simple voice commands. However, these powerful functions cannot be "fitted" by a small speaker. In fact, Alexa 's AI functions are implemented by the Amazon's cloud computing center. The Echo speaker, as its name suggests, is just a microphone for the cloud computing center.

But when all computing is moved to the cloud, unexpected risks occur.

On March 2, 2018, many people found that Alexa was unresponsive when they tried to command Alexa as usual. The cause of the incident was that Amazon 's cloud service experienced a severe service outage that day. No matter what you said to it, Echo speakers and other Alexa devices would only respond with error messages.

This incident shows that when more and more functions are moved to the cloud, the purchased product is just an empty shell. Although it is usually powerful, once the remote end goes into trouble, the local end will be useless. Behind the advantages of the cloud computing mode, there are some insurmountable flaws. When the cloud computing center or network cannot be accessed, the embedded device will lose its intelligence. Specifically:

1. Cloud computing needs to be accessed through a remote network. Although with the development of wired and wireless broadband technology, the network seems to be everywhere, but it is still not accessible at anytime and anywhere. This is a problem for people who are always mobile or need to enter. For embedded devices in no man's land, dangerous areas, and unfamiliar worlds, you can't always rely on them. For example, for military robots, the wireless network is not only unreliable, but can even be destroyed by enemies. In addition, remote network access will cause delays and jitters, resulting in insufficient real-time response, which is fatal for some critical real-time processing tasks, such as car

driving. Cars must respond to road conditions in a very short period. An additional delay of 10 milliseconds may cause life-threatening consequences. What's more, the delay of the Internet is not fixed, sometimes fast and sometimes slow, and the occasional jitter is not suitable for watching online videos. It doesn't matter, but car driving needs to be foolproof, and every task must be completed within a limited time.

2. The bandwidth of the cloud computing center will also become a bottleneck, especially when many video and audio streams need to be processed simultaneously. The bandwidth of a single channel of video and audio is no longer a problem for network terminals, but when thousands or hundreds of channels of video and audio are aggregated into the cloud computing center, it may cause network congestion. Imagine that there are one million vehicles that use cloud computing to realize license plate recognition when entering and exiting the parking lot. On the one hand, the cloud computing center needs to spend a huge amount of money to purchase bandwidth. On the other hand, due to the periodicity of vehicle parking, during peak hours Network "traffic jams" will lead to real-world traffic jams.

3. Cloud computing will bring the risk of privacy leakage in network transmission, content storage, and other aspects. In some scenarios with high security and privacy requirements, such as smart homes, we hope to use video to analyze unexpected scenes in the home. Remote care is implemented for the elderly and young children, but no one is willing to publish their own videos online and let the Internet monitor them all the time.

4. The total cost of cloud computing services becomes increasingly expensive under long-term and heavy use conditions. Initially, the hardware cost of embedded devices is very low, and the price of cloud services amortized to each device is relatively cheap per month or year. However, after years of use, the accumulated costs begin to exceed the cost savings of embedded devices, and these costs continue throughout the life of the device. Don't forget, many embedded devices are designed to work year-round, such as security cameras. In this way, the cost of the cloud computing mode is not advantageous.

All the above reasons show that in many cases, the cloud computing mode cannot meet the requirements of embedded computing in terms of reliability, economics, and security, which makes it necessary to explore ways to implement artificial intelligence inside embedded devices without (completely) relying on the cloud. Embedded artificial intelligence will enter a new stage.

1.4 From Cloud to Device: Local Mode

The next stage in AI development is to bring deep neural networks from the cloud into the physical world. This is due to the research progress of artificial intelligence in embedded software and hardware in recent years. While initial efforts will naturally focus on shrinking existing deep neural network models into the limited processor and memory space of embedded devices, future implementations will also be based on the growing processing power of embedded chips as well as AI-accelerated chips developed specifically for AI.

A series of advances in semiconductor process integration and algorithm development, embedded devices (including mobile devices) are gradually getting rid of the shackles of the cloud and can independently perform some "heavyweight" tasks, such as automatic image tagging, biometric identification, and Robotic controls and the ability to perform them repeatedly and efficiently and instantly. This opens the door to embedded artificial intelligence.

Not surprisingly, embedded AI first made a breakthrough on the high-end embedded device: the iPhone. In 2017, the launch of Face ID facial recognition technology marked the beginning of the second phase of embedded artificial intelligence.

Functions such as smart voice assistants and face unlocking have gradually become standard features in consumer devices such as mobile phones and smart watches, indicating that AI will accelerate its penetration into daily life. But what confuses people is that most AI implementations on devices still use the cloud computing mode. These products act like puppets and sounding boards, with the real computing happening behind the scenes on cloud computing servers. Although very convenient, this implementation method violates the user's privacy.

Face ID is a biometric security system powered by a series of sensors and a new AI acceleration chip that uses a front-facing infrared camera to project 30,000 points to create an infrared and three-dimensional image of the user's face. It is accelerated by the Bionic Neural Engine of A11 and above models, which uses a dual-core design to perform up to 600 billion operations per second, enabling real-time processing. The A11 Bionic Neural Engine is not a general-purpose GPU. It is designed for specific neural network algorithms and supports Face ID, Animoji, photo tagging, and Siri voice assistant.

In a paper titled "An On-device Deep Neural Network for Face Detection "(Apple Inc., 2017a, b), Apple researchers describe how to implement the Face ID function based on the A11 Bionic neural network engine.

In 2017, when Apple researchers first started using deep neural networks for face detection in iOS 10, they realized that even the most high-end phones at the time were struggling to run deep neural network algorithms. Like other institutions, Apple had previously been using cloud-based systems for image recognition. To increase user privacy, image recognition algorithms are required to run on the device.

This article describes how Apple works within the confines of limited memory and CPU resources without interrupting other OS tasks and using a lot of extra power. The article details the technical details of how Apple adapts deep neural network models on an SoC-sized GPU. The A11 chip converts three-dimensional and infrared images into a mathematical representation and compares that expression with registered facial data to identify whether the person is using the iPhone. The article concludes

> *Combined, all these strategies ensure that our users can enjoy local, low-latency, private deep neural network inference without knowing that their phones are running neural networks at hundreds of billions of floating-point operations per second.*

In other words, the face recognition function of Face ID is implemented locally on the iPhone, rather than relying on cloud computing as before.

The iPhone also uses deep neural networks to recognize and analyze voice commands for Siri functionality. To do this, the iPhone uses an always-on, low-power auxiliary processor (AOP) to trigger Siri once it hears the user's "Hi, Siri" command, the AOP will wake up the main processor to analyze the user's voice with a more powerful deep neural network. As shown in Fig. 1.3. The benefit of this approach is that it requires minimal processing to listen for and detect the "wake word," saving valuable battery power on your iPhone, and once you wake up, you can take full advantage of the Bionic Neural Engine's powerful processing power. Of course, Siri has not completely escaped the shackles of the cloud, and complex multi-round interactive voice conversations are still handled by the server. This kind of model can be regarded as a device-cloud collaborative embedded artificial intelligence, which will be described in the following chapters.

Application developers can also use the neural network acceleration capabilities of iPhone hardware through the API and development tools of Apple's *Core* ML

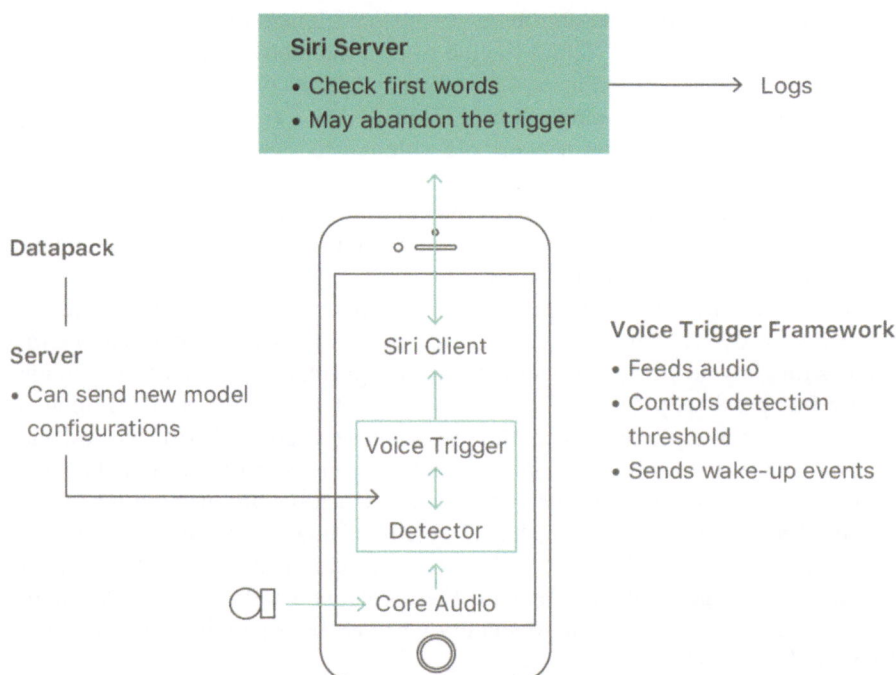

Fig. 1.3   Siri speech recognition process (Source: Apple) (Apple Inc., 2017a, b)

machine learning *framework*. As its tutorial details, the app can perform tasks such as shape recognition and object recognition.

Of course, Apple isn't the only player in this space. ARM, Google, Microsoft, and other companies have also begun to introduce AI into embedded devices.

1.4.1   ARM

ARM is used in most mobile devices and is also the developer of App authorized and customized processor platforms. It introduces AI into its universal SoC design, which will greatly expand the popularity of AI-accelerated devices.

The design, called DynamicIQ, adds processor instructions designed to accelerate machine/deep neural network algorithms, and ARM expects to improve AI performance by 50 times over the next 3–5 years relative to current ARM systems. Some companies are already using low-power ARM-M processors for embedded artificial intelligence applications. For example, the Amiko Respiro, an inhaler for asthma patients, uses data from multiple sensors and onboard machine learning software to calculate the drug's effectiveness and develop therapies customized for each patient.

1.4.2   Google

Not to be outdone, Google launches TensorFlow Lite platform that paves the way for deep neural network algorithms on mobile and embedded devices, TensorFlow Lite is designed to quickly launch TensorFlow models to fit into the small memory spaces of mobile devices and take advantage of any acceleration hardware, like embedded GPUs. The development framework also has interfaces to automatically use hardware accelerators on the device when available.

1.4.3   Microsoft

Microsoft is also developing embedded machine learning software for mobile and IoT devices, including the Raspberry Pi. This research currently focuses on narrow applications in specific scenarios, such as embedded medical devices or smart industrial sensors.

Other companies have also launched their solutions. For example, Reality AI provides a machine learning software library designed for embedded sensors and devices, which allows hardware devices with small physical size and harsh working environments to support more complex and accurate AI model.

Fig. 1.4 Local mode for
embedded artificial
intelligence

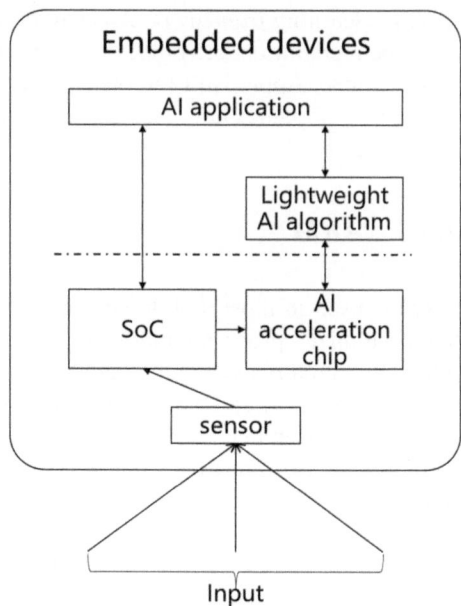

This series of progress has opened the second stage of embedded artificial intelligence. In this stage, AI hardware, algorithms and applications begin to get rid of the shackles of the cloud and move down to the embedded device itself. We call this local mode for embedded device. As shown in Fig. 1.4:

In this model, at the hardware level, an embedded AI acceleration chip is introduced, which has the characteristics of small size, low-power consumption, and high performance, and is specifically responsible for the inference operations of neural networks. At the software level, lightweight AI algorithms are introduced. These algorithms are improvements to traditional AI algorithms. Under the premise of completing the same function and approximate accuracy, the model has fewer parameters, so it takes up less storage space and has less computational complexity. It is small enough to be "loaded" into an embedded AI acceleration chip. Based on AI acceleration chips and lightweight AI algorithms, AI applications can be implemented locally on embedded devices, processing input signals obtained from sensors nearby, and achieving real-time calculation and response.

1.5 Technical Challenges of Embedded Artificial Intelligence

Although some breakthroughs have been made, embedded artificial intelligence still faces many technical challenges before large-scale application. At present, artificial intelligence is characterized by being computationally intensive, memory intensive, data intensive, and energy intensive. The deployment cost is very high.

Taking AlphaGo as an example, it uses 1920 CPUs and 280 GPUs, and each game costs $ 3000 in electricity bills. Such luxurious configuration is difficult to achieve in embedded devices. But at the same time, we see that its human opponents, whether he is Lee Sedol or Ke Jie, have about 100 billion neurons in their brains, but the power is only about 20 W, which is accommodated in a small space of about 1.6 L (compared to computers), and there is no "God" in the cloud to support them. This sets a benchmark for the development of embedded artificial intelligence.

On the other hand, with the advancement of deep neural networks, the most advanced models are becoming more and more accurate, and the gap with human intelligence is getting smaller and smaller. But they are also becoming more and more complex, requiring higher and higher computing power and storage capacity.

This means that if the AI chip can achieve near-human energy efficiency and at the same time be able to load an artificial neural network close to human complexity into it, it will be just around the corner to achieve true human-like intelligence.

To catch up with this goal, embedded neural networks still face a series of technical challenges. The key is to develop a small but powerful embedded computing platform with strong enough computing power, low enough energy consumption, small enough memory footprint, fast enough speed, small enough volume, and light enough weight.

1.5.1 Model Size

The challenges faced by neural networks in embedded systems are largely caused by the size of the model. Assume the input image has dimensions 224 × 224. The size required for currently common image classification neural networks is given in Table 1.1:

In general, with the deepening of the network layer, the accuracy of the network (measured by Top −5 accuracy) is getting higher and higher, but the computational complexity of the network is getting stronger and stronger, and the number of parameters and memory usage are also increasing.

Table 1.1 Size of common image classification neural network models

Model	depth	Parameters (M)	FLOPS(G)	Size (MB)	Top-5 accuracy (%)
AlexNet	7	58.3	0.725	217	19.2
VGG16	16	134.2	15.5	512.2	9.9
ResNet18	18	11.7	1.814	23	10.92
GoogleNet	22	6	1.57	40	12.9
ResNet34	34	21.8	3.664	35	8.58
ResNetV1 50	50	25.6	3.858	97.7	7.7
ResNetV1 101	101	45	7.57	155	7
ResNetV1 152	152	60	11.3	230	6.7

Computational complexity is usually measured in FLOPS (floating-point operations per second), with orders of magnitude in GFLOPS (1 billion floating-point operations per second), TFLOPS (1 trillion floating-point operations per second), and so on. In embedded devices, computational complexity is usually measured in OPS (operations per second), with orders of magnitude units such as GOPS and TOPS. When an operation refers specifically to floating-point operations, there is no difference between the two units, FLOPS and OPS.

As the size of the input image increases, the required computational complexity, number of parameters, and memory footprint are also proportionally larger. Based on the current resolution of 1920 × 1080 commonly used in image monitoring, the computational complexity of the above model is between 100GOPS and 1TOPS, the number of parameters is between 200 M and 5000 M, and the storage space is between 1G and 20G.Models such as target detection and semantic segmentation are based on image recognition models, and their model sizes are relatively larger. Such a large-scale model is a huge challenge for embedded devices. There are currently some AI acceleration chips designed specifically for embedded environments that have achieved performance above 1 TOPS, but they still face some of the following challenges.

1.5.2 Energy Efficiency

To reduce the overall power consumption of embedded devices, the power of embedded chips is usually calculated in milliwatts, while the power of current mainstream GPU/TPU often exceeds 100 watts. This leaves embedded devices generally lacking the ability to perform deep reasoning for real-world applications.

From the perspective of energy efficiency, the energy efficiency of most AI chips that can be used in embedded devices is only 10–100GOPS/W, and only when the system efficiency is far more than 10TOPS/W can the always-on embedded neural network inference be truly realized. To this end, several improved GPU and neural processing unit (NPU) ASICs have recently emerged with energy efficiencies reaching and exceeding 1 TOPS/W. However, these systems are more suitable for scenarios where deep neural network inference is performed occasionally, such as on smartphones. However, in more typical scenarios, such as security cameras that need to be online 24 hours a day, they are still not efficient enough to keep real-time neural network processing in the milliwatt-level power range.

Figure 1.5 shows this clearly, with circles showing the energy efficiency required for several generations of neural network architectures to run in real time at 30 FPS.

With the advancement of neural network algorithms, the energy efficiency required by the latest model has dropped compared to the previous model and has dropped by two orders of magnitude compared with the original model. The triangle shows the energy efficiency required to execute these models on mobile devices, which is two orders of magnitude less energy efficient than always-on. The horizontal lines represent the energy efficiency that different types of AI chips can achieve.

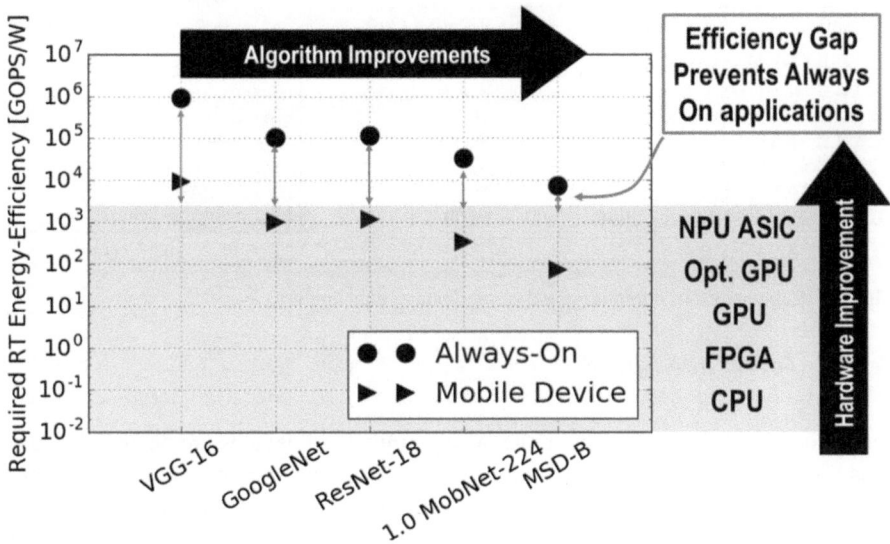

Fig. 1.5 Gap between energy efficiency requirements for embedded inference and existing capabilities (Moons et al., 2018)

The gap between the dots and the horizontal line represents the gap between demand and reality. This shows that although deep neural network algorithms and hardware have advanced by leaps and bounds in the past few years, and the latest innovations have indeed brought hardware platforms within the 100-mW power budget, there is still a considerable gap between them and the needs of embedded applications.

1.5.3   Memory Access

Complex deep neural network models tend to be measured in gigabytes in size, and as memory miniaturization and cost reduction, embedded devices have more than enough to store these models. However, more memory accesses mean greater energy consumption. As shown in Fig. 1.6, the power consumption of memory access is two orders of magnitude higher than that of arithmetic operations.

These models need to be read into the memory before inference in the embedded device. During the inference process, these memories are read repeatedly, which often means power consumption exceeding the power of the embedded device. This is completely unacceptable for some embedded devices that must use batteries. For embedded devices connected to the power grid, even if the power can be reduced by extending the inference time, it means greater total energy consumption. Overall, excessive memory usage remains a huge challenge.

Operation	Energy [pJ]	Relative Cost
32 bit int ADD	0.1	1
32 bit float ADD	0.9	9
32 bit Register File	1	10
32 bit int MULT	3.1	31
32 bit float MULT	3.7	37
32 bit SRAM Cache	5	50
32 bit DRAM Memory	**640**	**6400**

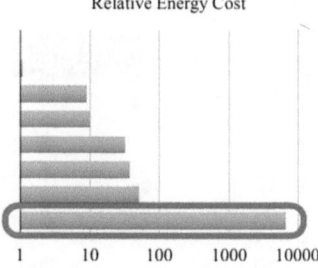

Fig. 1.6 Comparison of computing and storage power consumption

1.5.4 Inference Speed

Inference speed is another challenge. Many embedded devices are designed to process a steady stream of inputs, such as video surveillance cameras, and must carefully examine the possible targets in each frame, i.e., if the frame rate of the video is 25 FPS, it must complete one inference within a latency of 40 ms before completing the next inference. When we want to deploy deep neural networks in more demanding environments, such as video obstacle avoidance cameras on drones, the required inference speed will be higher, and the latency requirement will be at the 1 ms level. Such speeds are a challenge even for high-performance GPUs in servers.

1.5.5 Size and Weight

High-performance GPUs/TPUs are deployed in cloud computer rooms, which have relatively loose requirements for size and weight. However, for AI hardware to be deployed in embedded devices, if they are too large, they may not fit into the small body of the device, and the weight is too large, which can cause the overall energy consumption of the device to rise dramatically. For example, for a drone, although highly intelligent "eyes" are very important, if the eyes are larger than the "head" and the weight exceeds the lift, it may not even be able to take off. Of course, size and weight are positively related to the energy efficiency and total energy consumption of the hardware. Solving the problem of energy efficiency will, to a certain extent, also solve the problem of size and weight.

Although there are many conditions, the technical bottlenecks in embedded neural networks will eventually be overcome by following the example of the biological brain and the computation of the neural network of the human brain. This requires algorithms, hardware, and applications to be optimized and work closely together with each other. This book will take a deep dive into how to do it.

1.6   Approaches to Implementation Embedded Artificial Intelligence

So, how can complex artificial intelligence be implemented in embedded devices with limited resources? Let's look at some of the ways artificial intelligence can be applied on embedded devices:

1.6.1   Inference

In the computation of deep neural networks, training consumes a lot of computing power, while inference requires relatively low computing power. Simple inference tasks such as image classification, gesture recognition, speech detection, and motion analysis can be done on edge devices. Since only the final result is transmitted, latency is minimized, privacy is improved, and bandwidth is saved in IoT systems.

1.6.2   Hierarchical Inference

The inference of multi-layer neural networks can also be distributed to devices and the cloud, with lower-level reasoning performed on the device and higher-level reasoning performed in the cloud. This balances workload and latency. The first few layers of a neural network can be viewed as feature abstraction layers. As information propagates upward in the network, they are abstracted into high-level features. These advanced features take up much less storage space than the raw data and use less bandwidth to transmit over the network.

1.6.3   Transfer Learning

In the hierarchical inference above, the neural network is distributed between the device and the cloud. In this way, neural networks can be repurposed for completely different applications just by changing high-level layers in the cloud. Application logic in the cloud is easy to change. This hot swapping of neural network layers allows the same device to be used for different applications. This approach can be seen as an example of transfer learning: modifying a part of the network to perform a different task. As shown in Fig. 1.7.

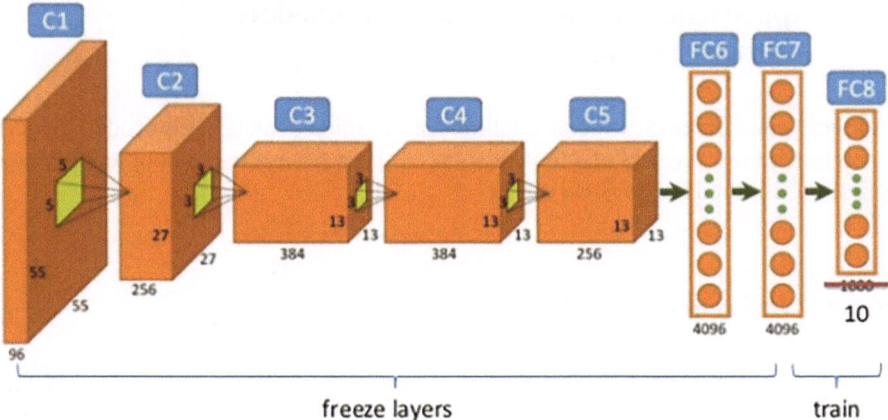

Fig. 1.7 Transfer learning (Neptune Labs, 2023)

1.6.4 Generated Model

Based on the above hierarchical learning and transfer learning, through careful design, an approximation of the raw data can be reconstructed from the extracted features. This allows edge devices to generate complex outputs with minimal input from the cloud, for example for data decompression.

1.6.5 Federated Learning

Once the equipment is deployed in the field, improvements can be made over time. Google's Gboard, for example, uses a technique called federated learning, as shown in Fig. 1.8: each device collects data and makes individual improvements, then aggregates those individual improvements on a central server, and finally updates each device with the combined results.

It may take time for low-power and low-cost AI hardware to become widespread. However, through some of the above methods, combined with the rapid evolution of deep neural network algorithms, the implementation of embedded neural networks can be expected in the future.

1.7 Components of Embedded Artificial Intelligence Implementation

Based on the methodology introduced in the previous section, we can start trying to develop artificial intelligence programs for embedded devices.

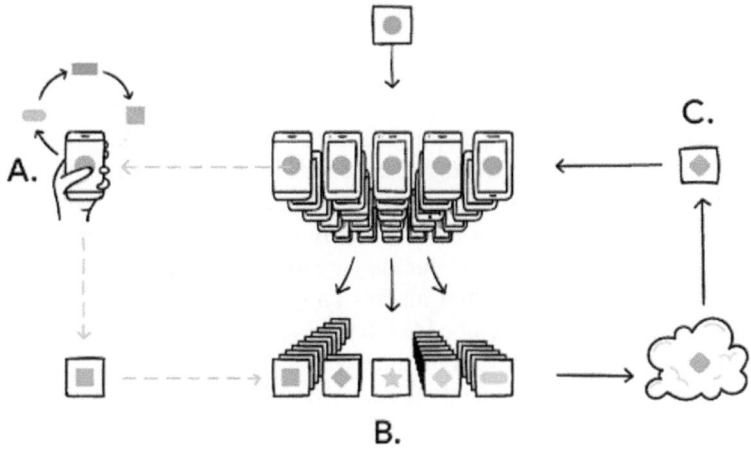

Fig. 1.8 Federated learning

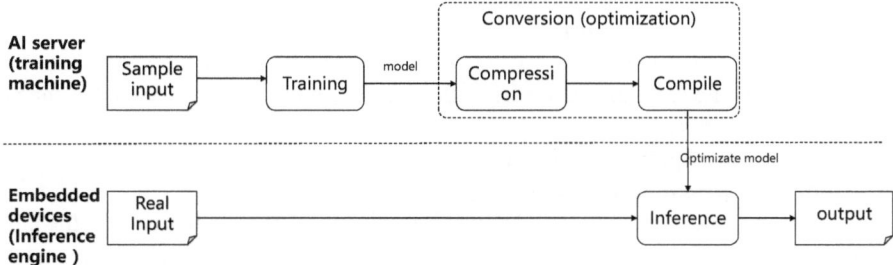

Fig. 1.9 Embedded artificial intelligence development process

In the traditional embedded development process, to make up for the lack of performance of the embedded device, the embedded software is not developed locally on the device, but with the help of the host. The host completes code writing, compilation, linking, debugging, etc., and then downloads the cross-compiled binary program to the target machine (embedded device) for testing and execution.

Similarly, the development of embedded artificial intelligence is not completed by embedded devices alone but is completed with the help of AI servers (training machines). The embedded devices are mainly responsible for inference, called inference machines. As shown in Fig. 1.9:

First, the model is trained on the AI server (training machine) with the help of its powerful AI hardware, such as GPU/TPU. This usually takes days or even weeks to generate a general model in some format that contains the structure, parameters, etc. of the neural network. There are often more than millions of parameters, occupying more than G of storage space. Then, the AI server converts (or optimizes) this general model into an optimized model designed to run on a certain embedded device.

Conversion (optimization) is divided into two steps, compression and compilation. Compression methods include pruning and quantization, etc., thereby reducing the number of parameters and storage occupation of the model. Compilation is to translate the model into computing instructions on the AI accelerator dedicated to embedded devices and perform targeted optimization to generate binary code. Finally, the converted model is downloaded to the embedded device, inference is performed based on the actual input, and the output is obtained.

This collaborative division of labor not only makes the development of embedded artificial intelligence possible, but also greatly reduces the pressure of deploying artificial intelligence algorithm models on embedded devices. Only inference needs to be performed on the embedded device, and the requirements for AI accelerators and models are significantly reduced.

Although with this development method, we still need to solve some of the core problems in the implementation of embedded artificial intelligence:

1. **Embedded AI chip**

   The AI chip is the carrier for deep neural networks to run. Only if it is powerful and compact enough can embedded artificial intelligence become possible.

2. **Lightweight AI algorithm**

   Undoubtedly, the smaller the size of the model, the lower the computational complexity, the less hardware performance requirements, and the easier it can be loaded into embedded devices. We need lightweight AI algorithms, which in most cases are lightweight neural networks.

3. **Model compression technology**

   If the architecture of the model itself remains the same, wouldn't it be the best of both worlds if the model compression technology can make the model smaller and achieve the same inference accuracy?

4. **Model compilation technology**

   An algorithm is composed of a series of operations, and these operations are ultimately completed on the AI chip. If the algorithm-compiled computing instructions are perfectly matched to the chip, the efficiency of embedded devices in performing AI tasks will be further improved.

5. **Embedded AI application framework**

   AI applications can cleverly use the local AI capabilities of embedded devices, and even call cloud AI capabilities when necessary to collaboratively complete tasks and maximize overall efficiency.

We collectively refer to the implementation of these five issues as the five major components of embedded artificial intelligence implementation, as shown in Fig. 1.10:

In the following chapters, we will introduce each of these components in depth.

Fig. 1.10 Five components of embedded artificial intelligence implementation

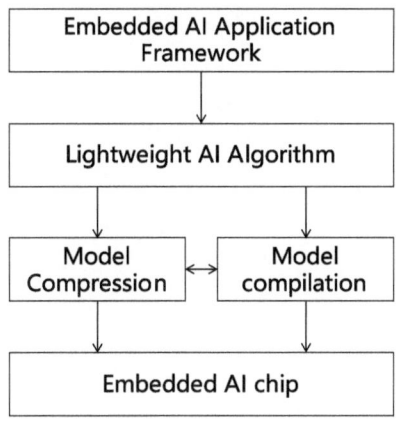

References

Apple Inc. (2017a). *An on-device deep neural network for face detection.* Retrieved from https://machinelearning.apple.com/research/face-detection

Apple Inc. (2017b). *Hey Siri: an on-device DNN-powered voice trigger for Apple's personal assistant.* Retrieved from https://machinelearning.apple.com/research/hey-siri

Moons, B., Bankman, D., & Verhelst, M. (2018). *Embedded deep learning: algorithms, architectures and circuits for always-on neural network processing 1st ed. 2019 Edition.* Springer.

Neptune Labs. (2023). *Transfer learning guide: a practical tutorial with examples for images and text in Keras.* Retrieved from https://neptune.ai/blog/transfer-learning-guide-examples-for-images-and-text-in-keras

Chapter 2
Principle of Embedded AI Chips

Abstract This chapter introduces the first component to implement embedded AI, the principle of the embedded AI chip. These chips can be streamlined versions of GPUs, TPUs, or ASICs and FPGAs designed for specific purposes. When needed, they will be integrated into embedded SoC chips. These chips adopt a parallel computing architecture and introduce concepts such as systolic arrays and multi-level caches to optimize data flow and minimize energy consumption by reducing memory access time during calculations. Multiple data flow strategies optimize data reuse and locality through innovative architectural approaches to reduce overall computing load and power requirements. This chapter also introduces the application of sparse matrix techniques that help compress data and speed up processing time.

Keywords Embedded AI chip · Parallel computing · Systolic array · Sparse inference

To deploy AI to embedded devices, it is first necessary to improve the hardware performance of embedded devices so that they are capable of at least simple inference tasks. In the cloud, to realize the training and inference of artificial intelligence, we generally use dedicated hardware, such as GPU/TPU/ASIC and FPGA, which are specially used to run deep neural network algorithms in addition to the CPU, and their performance is at least ten times higher than that of the CPU. Therefore, in the same way, in embedded devices, such hardware accelerator chips are also needed, which can work in tandem with the SoC and, in the future, can also be directly integrated into the SoC chip to enhance the AI processing power of the embedded device (Sze et al., 2020; Han et al., 2016).

2.1 Parallel Computing

To achieve high-performance inference, accelerator chips generally use parallel computing architectures.

© Tsinghua University Press 2024

B. Li, *Embedded Artificial Intelligence*,
https://doi.org/10.1007/978-981-97-5038-2_2

In neural networks, there are many matrix operations. For example, for the fully connected layer, you can use matrix multiplication, and for convolutional layers, you can use Toeplitz matrix multiplication. If the efficiency of matrix operations can be improved, the computation of neural networks can be accelerated.

In the process of matrix multiplication, multiplication and accumulation occur at the same time, that is, the product result of multiplication and the value of the accumulator are added and then stored in the accumulator, that is, $a = a + b \times c$. If multiplication and accumulation are divided into two operations, the intermediate result is stored in the register after the multiplication operation is completed, and then read out of the register during the addition operation. Not only does this require two clock cycles to complete, but it also adds additional read and write time, which is less efficient. So, in AI accelerators, they are unified into one operation: multiplication and addition (multiply-and-accumulate, abbreviated as MAC). In a matrix operation, multiple MAC operations need to be performed at the same time, and in the process of neural network calculation, multiple matrix operations need to be performed at the same time. If parallel processing can be achieved, the efficiency of the system will undoubtedly be greatly improved, so the AI accelerator adopts a parallel computing architecture, with two models: temporal concurrency and spatial concurrency, as shown in Fig. 2.1.

Regardless of the architecture, the processing array is composed of processing elements (referred to as PE), and the core of PE is the arithmetic logic unit (referred to as ALU). Compared with the ALU in the CPU, the ALU of the AI acceleration

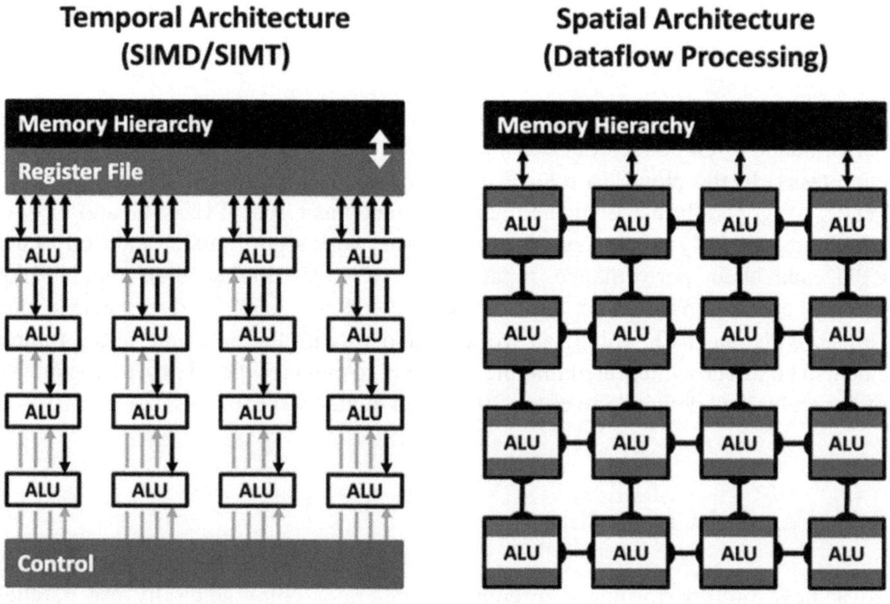

Fig. 2.1 Parallel computing architecture for AI accelerators (Source: arXiv:1703.09039) (Sze et al., 2020)

chip is very simple, mainly multiplying and adding, so the implementation structure is simple, the cost is low, and it is convenient to form a large-scale array.

Based on processing array, the following methods can be used to further improve the efficiency of MAC computing:

- Increase the speed of the processing element:
  Methods such as speeding up the clock frequency increase the processing power of a single processing element.
- Increase parallelism:
  That is, to increase the number of processing elements on the chip, so that more MAC operations can be processed in parallel.
- A single instruction implements more MAC operations:
  This requires parallel computing, and in GPUs, SIMT (Single Instruction, Multiple Threads) technology is used to broadcast an instruction to multiple processing elements to achieve parallelism. For example, 64 MAC addresses are encapsulated into a new instruction, MMA (Matrix Multiply Accumulate).
- More MAC operations in a single clock cycle:
  Either increase the memory bandwidth or reduce the accuracy of the data. For example, with a memory bandwidth of 512 bits, it can perform 16 32-bit MAC operations or 64 8-bit MAC operations. Doubling the memory bandwidth, or halving the data accuracy, can double the number of MAC operations in a single clock cycle.

Parallel computing is the foundation of AI accelerators. How to make good use of the processing element array requires the following technologies.

2.2 Systolic Array

In the computation process of neural networks, memory access is the biggest bottleneck. When the ALU performs multiplication and addition, it needs to read and write to the memory, read the weight value and activation value, and write the calculation result to the memory, as shown in Fig. 2.2.

Memory access is more energy-intensive than arithmetic operations, so the key to achieving energy efficiency in AI accelerators is to minimize memory access. Taking AlexNet as an example, an inference process requires 724 M times

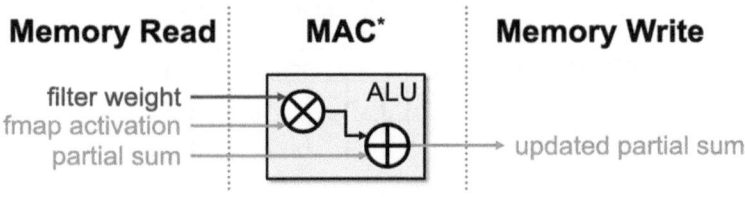

Fig. 2.2 Memory reading and writing during ALU multiplication and addition (Sze et al., 2020)

multiply-and-add operations, and if not optimized, this means 2896 M memory accesses, which is very inefficient.

To reduce the amount of memory read and write during computation, systolic array (Ross et al., 2017) are an effective solution. The logic of the systolic array is simple, since it takes more time to read the memory, the systolic array tries to run more computation in the process of a memory read to balance the time consumption between storage and computation.

The systolic array is shown in Fig. 2.3.

The upper part of the figure is a model of a traditional computing system. A processing element (PE) reads data from memory, processes it, and writes it back to memory. The biggest problem with this system is that the speed of data access is often much slower than the speed of data processing. As a result, the processing power of the entire system (MOPS, millions of operations per second) is largely limited by the ability to access memory. The systolic architecture uses a very simple approach: let the data flow in the processing element for as long as possible.

As depicted in the lower half of the figure above, the first data goes to the first PE, is processed, and passed to the next PE, and the second data goes to the first PE. And so on, when the first data reaches the last PE, it has already been processed multiple times. So, the systolic architecture is reusing the input data multiple times. As a result, it can achieve high computing throughput while consuming less memory bandwidth.

When a systolic array is to perform a neural network calculation, that is, perform a calculation of $Y = WX + b$, where the calculation of the systolic array is divided into three steps, the input is pixel X and the weight W, and the output is Y at one time. Inflow array with the weight W from top to bottom. Flows pixels X from left to right into the array. Each cell accepts only one piece of data in each interval from left and from top, multiplying them and accumulating them into the previous results until no

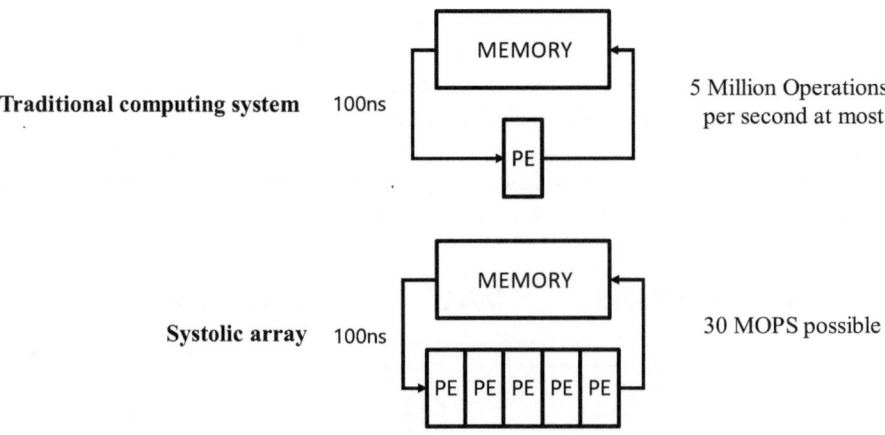

Fig. 2.3 Systolic array

data flows in. When the "systolic" process is over, Y is also calculated, as shown in Fig. 2.4.

The above design can not only maximize data reuse, reduce the number of memory visits of the chip during the computing process, but also reduce the pressure on memory bandwidth, thereby reducing the energy consumption of memory access. As a result, the systolic array is capable of processing hundreds of thousands of matrix operations in a single clock cycle.

2.3   Multi-Level Cache

Another way to reduce memory reading and writing is to cache, that is, to add cache before DRAM reading and writing, which are smaller but faster and more energy-efficient, as shown in Fig. 2.5.

With the introduction of caching, the cache enables data reuse before reading DRAM, and local accumulation before writing DRAM.

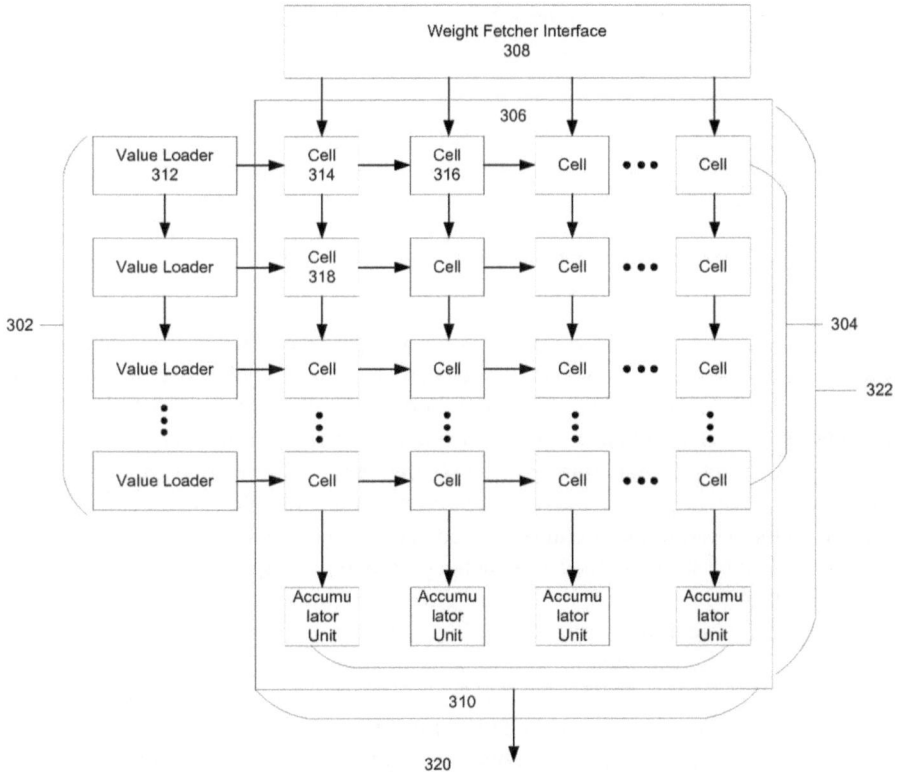

Fig. 2.4 Systolic array performs neural network calculations (Source: Google patent US9747546B2)

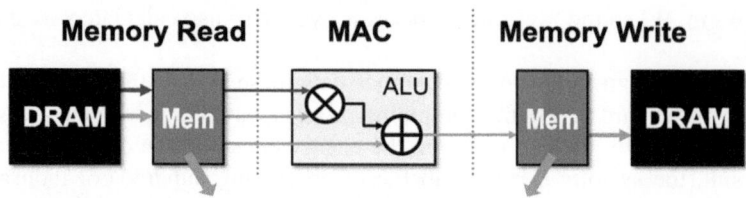

Fig. 2.5 ALU caching

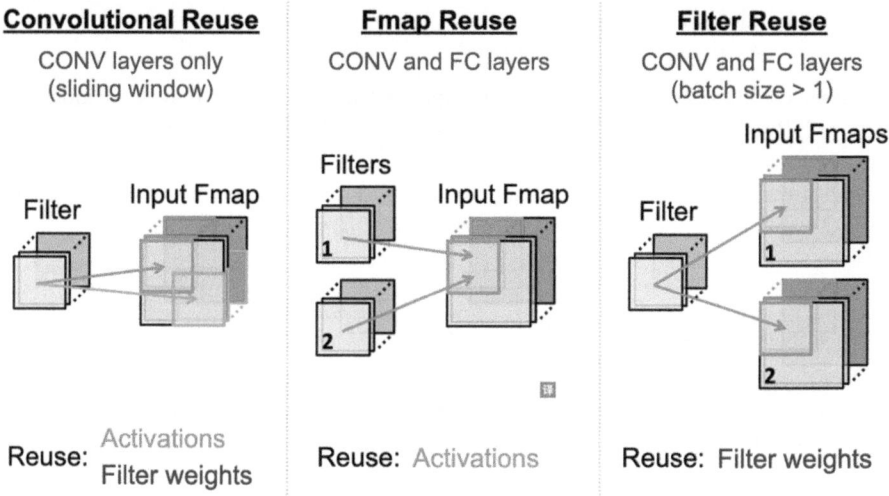

Fig. 2.6 Parameter reuse in convolutional neural networks

The characteristics of convolution operations determine that the weight and activation values of the convolution kernel can be reused repeatedly. As shown in Fig. 2.6, when convolution is performed on the same feature map, different sliding windows share the weight and activation values; The convolution kernel is also used repeatedly on multiple feature maps, so that activation values can be shared; Whereas, between multiple layers, sometimes the convolutional kernel is also shared, so the weight values can be shared. These shared weight values and activation values can be stored in the LU's cache, greatly reducing direct access to DRAM.

For example, in AlexNet, in the best case, the number of reads of the convolutional kernel and feature map is reduced by a factor of 500 due to data reuse, and at the same time, the accumulation results of partial sums do not require access to DRAM due to local accumulation, and in general, this caching mechanism reduces the number of memory visits from 2896 M to 61 M.

All concurrent processing elements can share a global cache, and there are also caches between adjacent processing elements and within the processing elements

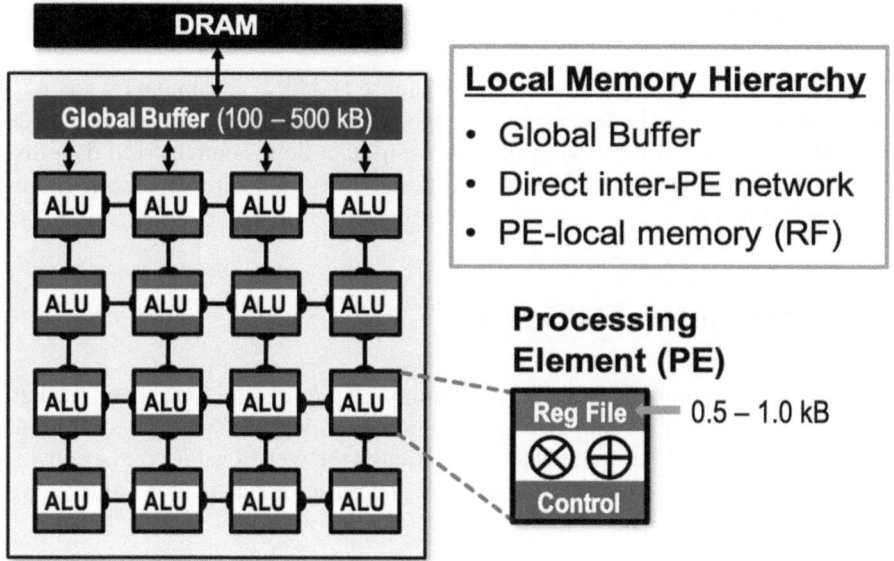

Fig. 2.7 ALU cache hierarchy

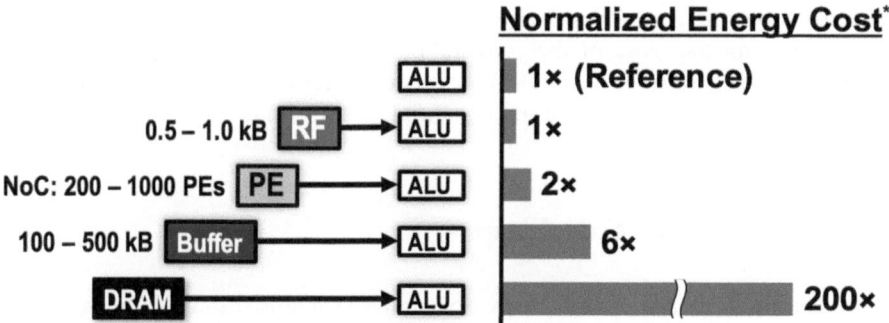

Fig. 2.8 Schematic diagram of energy efficiency of the cache hierarchy

(Reg File, abbreviated as RF), thus forming a hierarchical cache structure, as shown in Fig. 2.7.

   Among them, the cache inside the processing element is the smallest, the cache between the processing elements is the second, and the global cache is the largest. Each layer of cache introduced can reduce the number of DRAM accesses exponentially, which adds up to about 200 times the total energy efficiency, as shown in Fig. 2.8.

2.4 Data Flow

However, caching alone is not enough, data reuse and local accumulation are maximized at low cost and high concurrency, and special processing data flows are also required. There are several types of processing data flows: output fixed data flow, weight fixed data flow, input fixed data flow, and row-fixed data flow. Each of these is described below.

2.4.1 Output Fixed Data Flow

This data flow minimizes the energy consumption required for partial and read and write operations by maximizing local accumulation. When implemented, it broadcasts or multicasts the weight values or multicasts the weight values of the convolution kernel on the processing element array and reuses the activation values across the spatial latitude, as shown in Fig. 2.9.

2.4.2 Weight Fixed Data Flow

This data flow minimizes the energy consumption required to read the weight values by maximizing the reuse of the convolution kernel weight values. When implemented, it broadcasts the activation value on the processing element array and accumulates the partial sum over the spatial latitude, as shown in Fig. 2.10.

2.4.3 Input Fixed Data Flow

This data flow minimizes the energy consumption required to read the activation values by maximizing the reuse of convolutions and feature map activation values. When implemented, it adds up the unicast weight values on the processing element array and the partial sum over the spatial latitudes, as shown in Fig. 2.11.

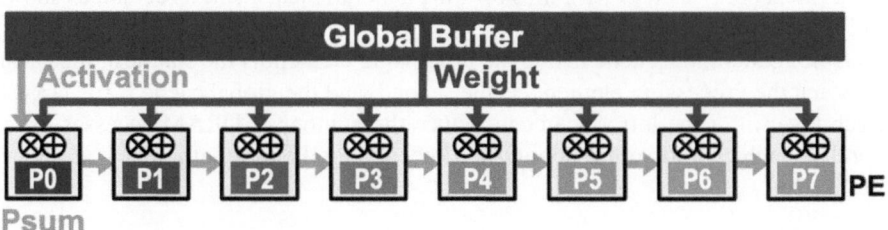

Fig. 2.9 Output fixed data flow

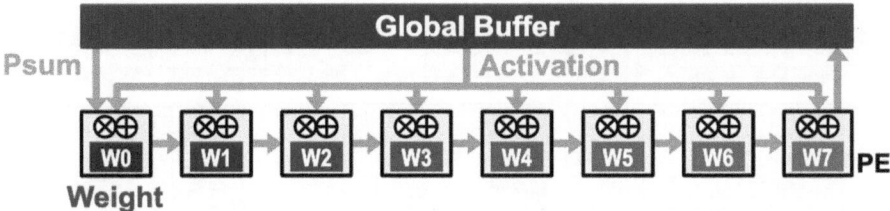

Fig. 2.10 Weight fixed data flow

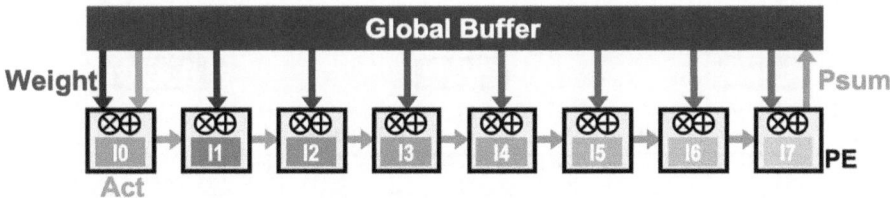

Fig. 2.11 Input fixed data flow

2.4.4   Row-Fixed Data Flow

This data flow enables overall energy efficiency optimization by maximizing data reuse and accumulation on the RF, rather than optimizing only for certain data types.

Using the one-dimensional convolution operation, the row-fixed data flow allocates the processing of the one-dimensional row-convolution to each PE. The weight values of a row in the convolutional kernel are kept fixed inside the RF of the PE, and the input activation values are then streamed into the PE. During the MAC address operation performed by PE for each sliding window, these MAC addresses use the same storage space to add up the partial sum. Since the input activation values are duplicated between the different sliding windows, the input activation values can be kept in the RF and reused. By traversing all the sliding windows, it completes one-dimensional convolution and maximizes data reuse and local accumulation, as shown in Fig. 2.12.

In two-dimensional convolution, you only need to arrange the PEs into a spatial array and allocate the rows of different convolution kernels to different PEs to run, which can achieve the same effect as 1D convolution, as shown in Fig. 2.13.

In general, these data flows minimize data movement from different aspects. The output fixed data flow minimizes the movement of the partial sum, the weight fixed data flow minimizes the movement of the weight values, the input fixed data flow minimizes the movement of the input values, and the row-fixed data flow minimizes the movement of both the convolution kernel and the partial sum. The result is an

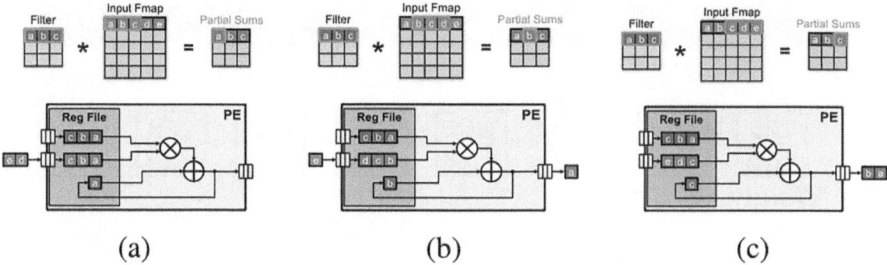

Fig. 2.12 Row-fixed data flow for 1D convolution. (**a**), (**b**), (**c**) are 3 steps of the data flow

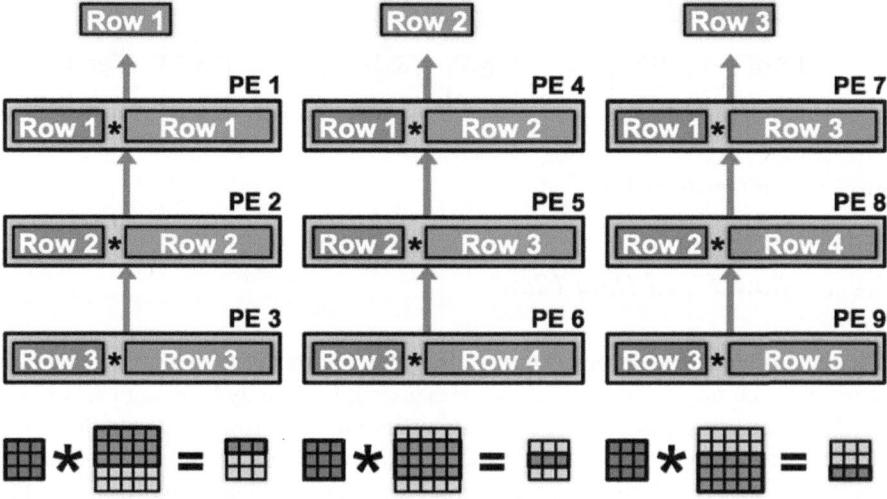

Fig. 2.13 Row-fixed data flows for 2D convolution

increase in energy efficiency. Among them, the row-fixed data stream is the most efficient, 1.4 to 2.5 times more efficient than other data flows.

Figure 2.14 shows the decomposition of energy consumption for each layer in AlexNet after applying the row-fixed data flow technique. In the convolutional layer, the energy is mainly consumed by the cached RF inside the PE, while in the fully connected layer, the energy is mainly consumed by DRAM. In total, the convolutional layer consumes about 80% of the energy, accounting for the majority. As the latest deep neural network model layers become deeper and deeper, the more convolution layers there are, the ratio between the number of convolutional layers and fully connected layers will only become large and larger, and the more obvious effect of the fixed data flow technology become.

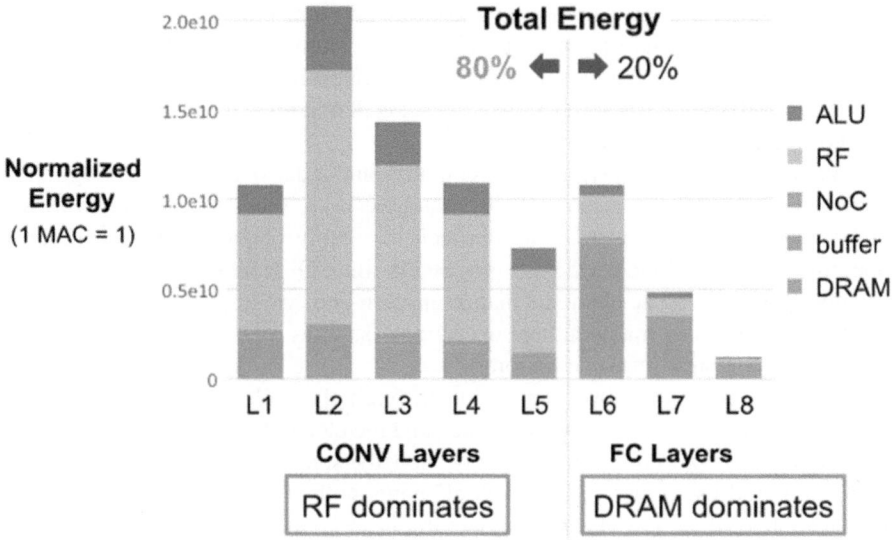

Fig. 2.14 Distribution of energy consumption in each layer after AlexNet applies fixed data flow technology

2.5   Sparse Inference

First introduce the concept of sparse matrix.

We know that a matrix is a two-dimensional data object consisting of M rows and N columns, so there are $M \times N$ values in total. When most of the values in this matrix are zero and the non-zero elements are irregularly distributed, the matrix is called a sparse matrix.

Sparse matrices are very common in machine learning. Although they arise naturally during some data collection processes, more often they are obtained when specific data transformation techniques are used. For example, we use an algorithm to extract part-of-speech tags as supplementary tags to train a sequence classifier; if extracted in the entire document set, the result can be vectorized into a very large sparse matrix and passed to the classifier as a parameter.

Since sparse matrices contain many zero-valued elements, we can use specific algorithms to do two important things:

1. Compress the memory footprint of matrix objects.
2. Accelerate the computation speed of machine learning.

Storing a sparse matrix requires allocating memory for each byte or even 64-bit zero value in the matrix, which is an obvious waste of memory resources because these zero values do not contain any information. We can use compression techniques to minimize the amount of data we need to store. That's not the only benefit.

Almost all-natural machine learning algorithms require the data matrix to exist in memory in advance, that is, when the data matrix cannot be completely stored in the memory, the machine learning process will be interrupted. One of the benefits of converting a dense matrix into a sparse matrix is that in most cases the sparse matrix can be compressed to fit in memory.

Again, consider multiplying a sparse matrix and a dense matrix. Although there are many zero values in a sparse matrix, and we know that zero multiplied by any number is zero, the conventional approach inevitably performs this meaningless operation. This significantly delays processing time. Obviously, it is more efficient to only operate on those elements that return non-zero values. Therefore, any algorithm that uses basic mathematical operations (such as multiplication) can benefit from a sparse matrix implementation.

We can verify the above conclusion through the following simple example. Suppose we have a $2000 \times 10,000$ data set, and the elements of this matrix have only two values 1 and 0. In order to intuitively show the sparsity of this matrix, we create a quadrant and use points to represent elements with a value of 1, while elements with a value of 0 are left blank, as shown in Fig. 2.15.

As you can see, the imaged data set is mostly blank. For comparison, the Fig. 2.16 shows a dense matrix with the same number of elements.

If you convert a dense matrix into a sparse matrix, the size of the dense matrix is 160 MB, while the size of the sparse matrix is 24 MB, which is equivalent to a compression rate of 85%!

Next let's look at the comparison in terms of computation time. Test three different classification algorithms: Bernoulli Naive Bayes, Logistic Regression, and Support Vector Machines. The naive Bayes classifier runs eight times faster under

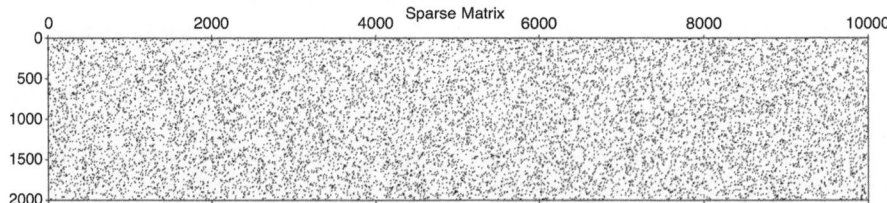

Fig. 2.15 Sparse matrix

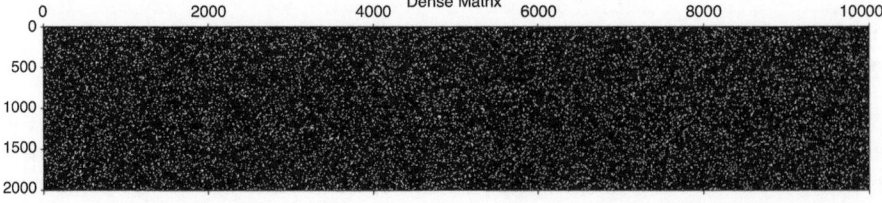

Fig. 2.16 Dense matrix

sparse matrices; for logistic regression, the processing time is reduced by about 33%; for support vector machines, sparse matrices only use half the processing time compared to dense matrices! In general, converting a dense matrix into a sparse matrix form almost always improves processing time.

To obtain the above benefits, sparse matrix algorithms are required.

Sparse matrix algorithm is the algorithm that uses sparse matrix as the core data structure. Compared with dense matrix algorithms, the biggest feature of sparse matrix algorithms is that it greatly reduces storage space requirements and computational complexity by only storing and processing non-zero elements. The price is that a special sparse matrix compression storage data structure must be used, so in many discrete indirect addressing operations are introduced in the calculation process. The sparse matrix algorithm is a typical irregular algorithm. The calculation memory access ratio is very low, and the memory access trajectory during the calculation process is related to the sparse structure of the sparse matrix. It is difficult to explore the spatiotemporal locality in the calculation process. Therefore, in the traditional Cache-based algorithm Sparse matrix algorithms are computationally inefficient on processors. To improve the computational efficiency of sparse matrix algorithms, existing algorithms need to be improved from two aspects: sparse storage data structures and sparse matrix algorithms.

According to different application fields, sparse matrix algorithms are divided into two categories. One category is non-numerical computing algorithms, typically represented by graph search algorithms, including core algorithms such as breadth-first search; the other category is numerical computing algorithms, typically represented by sparse linear algorithms. Solving systems of equations, including core algorithms such as sparse matrix-vector multiplication, solving sparse trigonometric equations, and sparse matrix decomposition.

Sparse matrix algorithms can be implemented in software or hardware.

The parameters of most current AI models are sparse to a certain extent. If sparse matrix algorithms can be implemented in embedded AI chips, their reasoning performance will undoubtedly be greatly improved.

The benefits of sparsity seem straight forward. But there have long been three challenges to realizing the promised gains.

Acceleration—Fine-grained, unstructured, weight sparsity lacks structure and cannot use the vector and matrix instructions available in efficient hardware to accelerate common network operations. Standard sparse formats are inefficient for all but high sparsity.

Accuracy—To achieve a useful speedup with fine-grained, unstructured sparsity, the network must be made sparse, which often causes accuracy loss. Alternate pruning methods that attempt to make acceleration easier, such as coarse-grained pruning that removes blocks of weights, channels, or entire layers, can run into accuracy trouble even sooner. This limits the potential performance benefit.

Workflow—Much of the current research in network pruning serves as useful existence proofs. It has been shown that network A can achieve Sparsity X. The

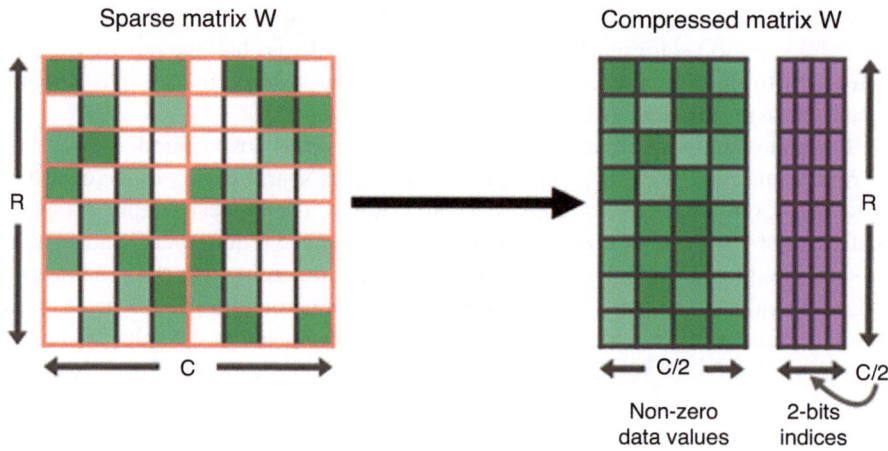

Fig. 2.17 Compressed sparse matrix (Source: arXiv:2104.08378 Accelerating Sparse Deep Neural Networks)

trouble comes when you try to apply Sparsity X to network B. It may not work due to differences in the network, task, optimizer, or any hyperparameter.

As an example, NVIDIA Ampere Architecture GPUs adds support for fine-grained structured sparsity to its Tensor Cores (Mishra et al., 2021). Sparse Tensor Cores accelerate a 2:4 sparsity pattern. In each contiguous block of four values, two values must be zero. This naturally leads to a sparsity of 50%, which is fine-grained. There are no vector or block structures pruned together. Such a regular pattern is easy to compress and has a low metadata overhead (as shown in Fig. 2.17).

Sparse Tensor Cores accelerate this format by operating only on the non-zero values in the compressed matrix. They use the metadata that is stored with the non-zeros to pull only the necessary values from the other, uncompressed operand. So, for a sparsity of 2x, they can complete the same effective calculation in half the time.

In detail, Sparse Tensor Cores perform sparse matrix × dense matrix = dense matrix operation. Figure 2.18 shows how a 2:4 sparse GEMM operation is mapped to Tensor Cores. 50% sparsity on one of the operands halves the required multiply-and-add operations, resulting in (up to) a 2× performance increase over equivalent dense GEMMs.

The figure shows the mapping a M × N × K GEMM onto a Tensor Core. Dense matrix A, of size M × K, (left side) becomes M × K/2 (right side) after pruning with 2:4 sparsity. Sparse Tensor Core hardware selects only the elements from B that correspond to the non-zero values in A, skipping the unnecessary multiplications by zero. In both dense and sparse GEMMs, B and C are dense K × N and M × N matrices, respectively.

In summary, when deploying a neural network, it's useful to think about how the network could be made to run faster or take less space. A more efficient network can make better predictions in a limited time budget, react more quickly to unexpected

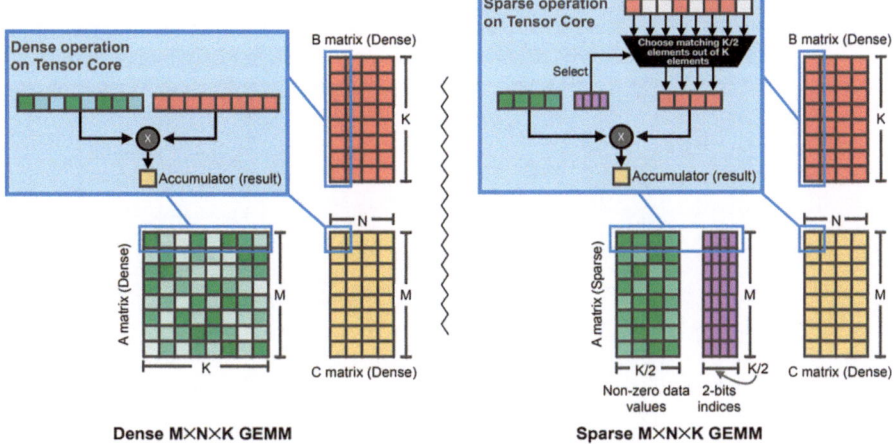

Fig. 2.18 Sparse Tensor Cores perform sparse matrix × dense matrix (Source: arXiv:2104.08378 Accelerating Sparse Deep Neural Networks)

input, or fit into constrained deployment environments. Sparsity is one optimization technique that holds the promise of meeting these goals. For embedded AI chips with limited computing power and memory, the role of sparse inference is more obvious.

Using the above technical principles, embedded AI hardware accelerators have developed rapidly in the past few years, and various forms of AI acceleration chips have emerged, some are lite versions of cloud GPUs, and some are FGPA, and ASICs designed for specific purposes. Although each has its own pros and cons, they are available options for accelerating edge AI inference.

References

Han, S., Liu, X., Mao, H., Pu, J., Pedram, A., Horowitz, M. A., & Dally, W. J. (2016). EIE: efficient inference engine on compressed deep neural network. arXiv:1602.01528.

Mishra, A., Latorre, J. A., Pool, J., Stosic, D., Stosic, D., Venkatesh, G., … Micikevicius, P. (2021). Accelerating sparse deep neural networks. arXiv:2104.08378.

Ross, J., Jouppi, N. P., Phelps, A. E., Young, R. C., Norrie, T., Thorson, G. M., & Luu, D. (2017). United States Patent No. US9747546B2.

Sze, V., Chen, Y.-H., Yang, T.-J., & Emer, J. (2020). *Efficient processing of deep neural networks*. Morgan & Claypool Publishers.

Chapter 3
Lightweight Neural Networks

Abstract This chapter introduces the second component to implement embedded AI, lightweight neural networks (abbreviated as lightweight network). Lightweight neural networks have small sizes and can operate within the constraints of low-power and memory-constrained environments while maintaining accuracy. Firstly, several strategies are introduced to reduce the computational complexity of neural networks without sacrificing accuracy. These strategies include the use of grouped convolutions, depthwise convolutions, pointwise convolutions, and depthwise separable convolutions. Next, some classic lightweight neural networks using the above strategies are introduced, such as SqueezeNet, Xception, MobileNet, Mnasnet, etc. By understanding the design ideas of these models, analyzing their network structures, and comparing the performance improvements they bring, we can learn how these models provide the best balance between performance and resource utilization, reducing parameter count and computational overhead. This chapter also lists some other common lightweight neural network models and introduces their application methods.

Keywords Lightweight network

The computing power and storage capacity of embedded systems are relatively limited, and it is best to use fast and efficient model architectures for neural networks to be applied to such low-power, memory-limited environments, especially if predictions are to be made frequently, such as identifying targets in live video.

Common neural networks, such as ResNet, will consume too much power and are large, making them unsuitable for real-time use.

In general, the larger the model, the better the results. However, the slower it runs, the larger the memory footprint and the more energy it consumes. Large models can exceed the limits of embedded systems in terms of memory and power, or quickly drain the battery of mobile devices. Smaller models run faster and have a smaller memory footprint, less power, and consume less battery on mobile devices but often produce less accurate results.

However, if the computational complexity of the neural network can be reduced, the size and power consumption of the model can be reduced while maintaining its accuracy, and such a lightweight neural network (abbreviated as lightweight network) may be applied to embedded systems or mobile devices.

© Tsinghua University Press 2024

B. Li, *Embedded Artificial Intelligence*,

https://doi.org/10.1007/978-981-97-5038-2_3

Lightweight neural networks are designed for traditional neural networks with deep model layers and a huge computational cost and parameters. These lightweight neural network models can be designed by humans, automated with the help of neural network architecture searches, or a combination of the two. In fact, this is how the best lightweight neural networks are currently available.

Some typical lightweight networks are introduced below. SqueezeNet, Xception, MobileNet, etc.

3.1 Reduce Computational Complexity

First, let's discuss general ways to reduce the computational complexity of neural networks.

The computational complexity of a neural network is mainly measured by the following two indicators.

1. Computational cost (FLOPS): the total number of floating-point operations of the model, and the total number of multiplication and addition operations in the statistical model. It determines the speed of network training and inference.
2. Number of parameters (params): the total number of weights and bias parameters of all layers of the model. The number of parameters is measured in the number of parameters, which determines the size of the memory occupied by the network. The actual memory size is multiplied by the number of bytes per parameter. For example, for FP32, each parameter occupies 4 bytes, while for FP16, it only occupies 2 bytes.

For a typical convolution operation, it is assumed that the input feature map size of the convolution layer is (H, W, C), the convolution tensor is (K, K, C, O), H is the height of the input feature degree, W is the width of the input feature map, C is the number of input channels, K is the size of the convolution kernel, and O is the number of output channels.

The computational cost of each convolution is: $K \times K$ multiplication operations (each parameter in the convolution kernel must be multiplied once by the elements on the feature map), and $K \times K - 1$ addition operations (the convolution result is, $K \times K$ numbers add up). Therefore, the number of multiplications and additions required for a convolution operation: $(K \times K) + (K \times K - 1) = 2 \times K \times K - 1$. In other words, it has a square relationship with the size of the convolution. The computational cost of the entire convolution layer is $H \times W \times C \times O \times (2 \times K \times K - 1)$. If the addition operation is ignored, it can be simplified to $H \times W \times C \times O \times K^2$. The number of parameters for each convolution layer is: weight parameter $K \times K \times C \times O$, bias parameter O, and the total number is $K \times K \times C \times O + O$. If the bias parameter is ignored, it can be simplified to $C \times O \times K^2$.

Obviously, if the number of convolution operations can be reduced, the computational complexity of the neural network will be significantly reduced. Methods to reduce the number of convolution operations are as follows.

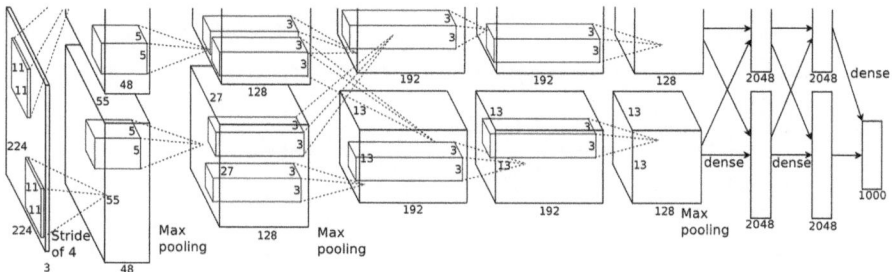

Fig. 3.1 Schematic diagram of AlexNet's use of grouped convolution (Krizhevsky et al., 2017)

3.1.1 Grouped Convolution

Grouped convolution is a variant of standard convolution, in which the input feature map is divided into G groups according to channels, and convolution is performed independently for each group, and the number of input and output channels of each group is the same as the original $1/G$, then the computational cost of each group convolution is $H \times W \times C/G \times O/G \times K^2$, *and* the computational cost of the total G group is $H \times W \times C \times O \times K^2/G$, which is Ungrouped $1/G$. The number of parameters also dropped to the original $1/G$, which is $C \times O \times K^2/G$. *Grouped* convolution originated from AlexNet. AlexNet won the championship in the ImageNet LSVRC-2012 Challenge, which is the pioneering work *of* convolution neural networks and leads a new round of development of artificial intelligence. AlexNet uses grouped convolution to divide the entire network into two groups, reducing memory requirements and training time, as shown in Fig. 3.1.

3.1.2 Depthwise Convolution

Depthwise convolution was first proposed by Google. It refers to dividing the input feature map into C groups according to channels. That is, depthwise convolution is a special simplified form of grouped convolution, and then each group is convolved separately. It is equivalent to collecting the spatial features of each input channel separately, as shown in Fig. 3.2.

In this way, the computational cost is reduced to $H \times W \times O \times K^2$ and the number of parameters is reduced to $O \times K^2$, both of which are $1/C$ of the ordinary convolution, which significantly reduces the computational cost and parameters.

The number of feature maps after depthwise convolution is the same as the number of channels in the input layer, and the feature map cannot be expanded. Moreover, this kind of operation independently convolution operation is performed on each channel of the input layer, and the feature information of different channels in the same spatial position is not effectively used. Therefore, pointwise convolution is required to combine these feature maps to generate a new feature map.

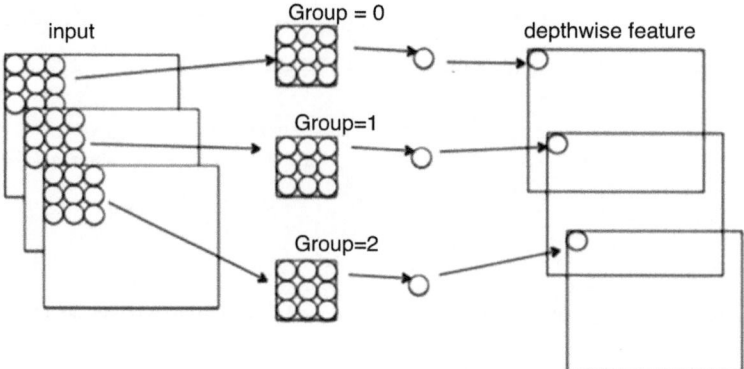

Fig. 3.2 Depthwise convolution

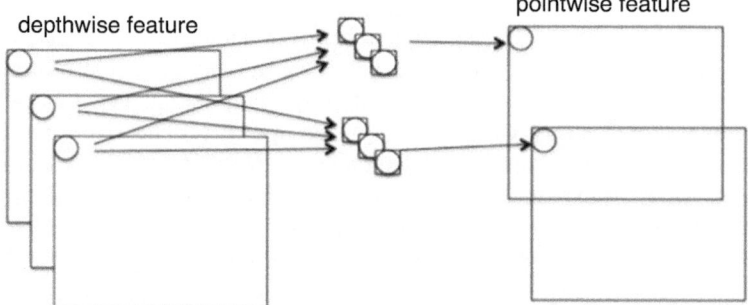

Fig. 3.3 Pointwise convolution

3.1.3 Pointwise Convolution

Pointwise convolution refers to the K ordinary 1×1 convolutions of input, which is used after the depthwise convolution, which is equivalent to mixing information between channels, as shown in Fig. 3.3.

The computational cost of pointwise convolution is $H \times W \times C \times O$, and the parameter quantity is $C \times O$.

3.1.4 Depthwise Separable Convolution

Depthwise Separable Convolution decomposes a complete convolution operation into two steps, i.e., depthwise convolution and pointwise convolution. It can complete the function of conventional convolution and extract features, but compared with conventional convolution operations, it has a lower computational cost and parameters.

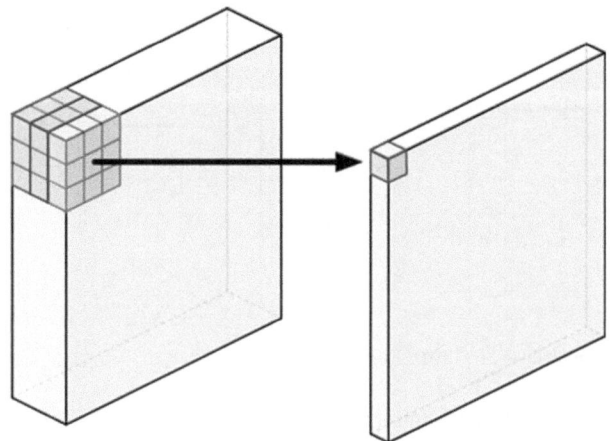

Fig. 3.4 Convention convolution diagram

The computational cost is $H \times W \times O \times K^2 + H \times W \times C \times O$, which is $1/C + 1/K^2$ of the conventional convolution.

The parameter quantity is $O \times K^2 + C \times O$, which is also $1/C + 1/K^2$ of the conventional convolution.

The following set of figures shows the difference between regular and depthwise separable convolution.

Regular convolution applies a convolution kernel (or "filter") to all channels of the input image. It slides this convolution kernel across the image and performs a weighted summation of the input pixels covered by the convolution kernel on all input channels at each step. The convolution operation combines the values of all input channels. If an image has three input channels, running a single convolution kernel on that image will result in an output image with only one channel per pixel, as shown in Fig. 3.4.

Depthwise separable convolution performs first depthwise convolution, with convolution operations performed separately for each channel, and then by performing pointwise convolution, adding up the outputs of each channel, as shown in Fig. 3.5.

3.1.5 Channel Shuffle Mixing

Grouped convolution causes the information flow of the model to be limited within each group, and there is no information exchange between groups, which will affect the representation ability of the model. Therefore, it is necessary to introduce a mechanism for information exchange between groups, that is, the channel shuffle operation. Channel shuffle was first proposed in ShuffleNet, which randomly

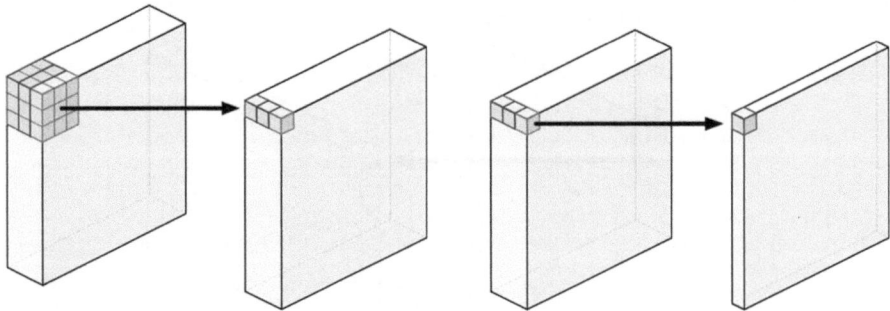

Fig. 3.5 Depthwise separable convolution diagram

connects input and output channels in convolution operations, changing the order between channels. It has no additional convolution calculations.

With the above general methods of reducing computational complexity, they can be applied to specific model designs.

3.2 SqueezeNet

SqueezeNet (Iandola et al., 2016) was released at the ICLR-2017 conference in 2017 by Berkeley and Stanford researchers.

3.2.1 *Core Idea*

SqueezeNet, which takes its name from Squeeze, introduces an extrusion layer that uses a 1 × 1 convolution kernel to convolute the feature map of the previous layer. The purpose is to reduce the number of channels in the feature map.

SqueezeNet is similar to AlexNet, but the number of parameters is only 1/50 of the former, and it uses model compression technology to compress the model size to 0.5 MB, which is only 1/510 of AlexNet.

SqueezeNet includes the following core ideas:

1. The number of parameters can be reduced by using a 1 × 1 convolution kernel instead of a 3 × 3 convolution kernel.
2. Limit the number of channels by squeeze the layer to reduce the number of parameters.
3. Referring to GoogleNet's Inception idea, parallel connection is performed after 1 × 1 and 3 × 3 convolutions. To make the feature maps have the same size, the 3 × 3 convolution kernel is padded.

4. Reduce the pooling layer and postpone the pooling operation as much as possible, so that the convolution layer has a larger activation layer, retains more information, and improves accuracy.
5. Use global average pooling instead of a fully connected layer.

The above 1–3 is implemented by the Fire module, as shown in Fig. 3.6 is the microarchitecture view of the Fire module, which shows the organization of the convolution filter in the Fire module. In this example, $s_{1 \times 1} = 3$, $e_{1 \times 1} = 4$, $e_{1 \times 1} = 4$. The convolution filters are shown here, but the activations are not shown.

The Fire module is mainly divided into two parts:

- Compression: 1×1 convolution kernel, parameter $s_{1 \times 1}$ represents the number of convolution kernels.
- Expansion: 1×1 convolution kernel and 3×3 convolution kernel, and the parameters $e_{1 \times 1}$ and $e_{3 \times 3}$, respectively, represent the number of two convolution kernels.

The module has a total of three parameters: $s_{1 \times 1}$, $e_{1 \times 1}$, $e_{3 \times 3}$, and the relationship remains $s_{1 \times 1} < e_{1 \times 1} + e_{3 \times 3}$.

3.2.2 Network Structure

The basic SqueezeNet consists of multiple Fire modules in series, and finally uses the global average pooling layer to output the results. Drawing on the idea of ResNet, bypass connections can be added between Fire modules to solve the problem of network degradation.

The network structure of SqueezeNet is shown in Fig. 3.7, with SqueezeNet on the left, SqueezeNet with simple bypass in the middle, and SqueezeNet with complex bypass on the right.

The parameters of each layer are given in Table 3.1.

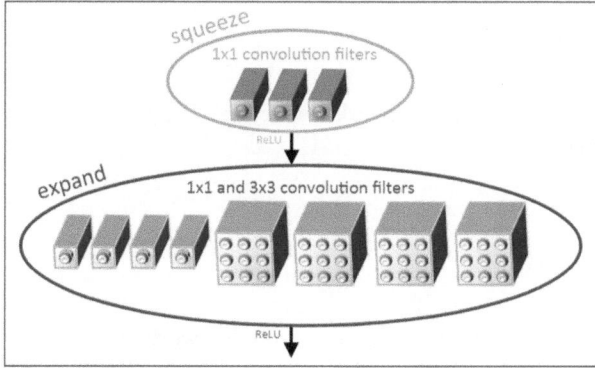

Fig. 3.6 Fire module of SqueezeNet (source: arXiv:1602.07360)

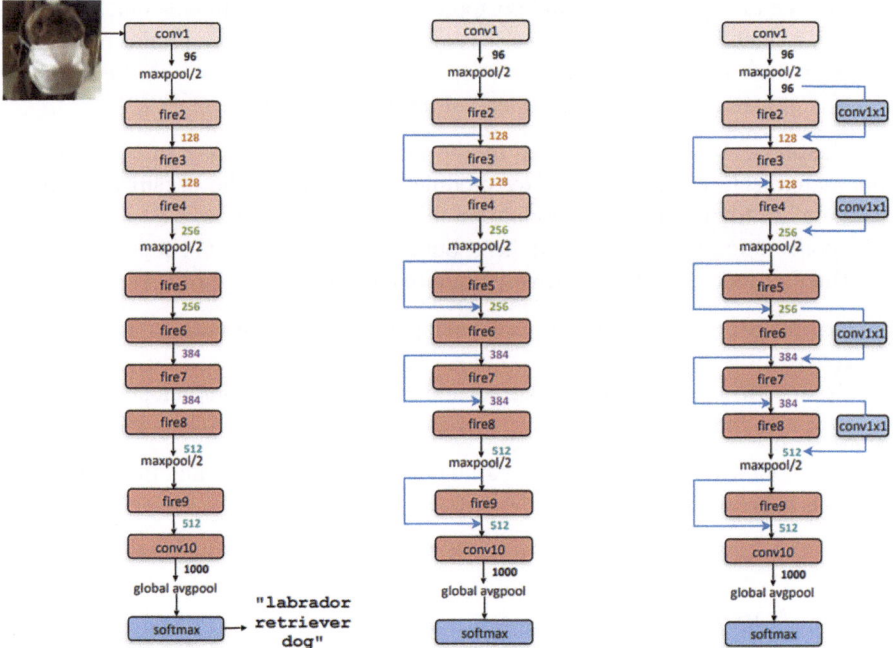

Fig. 3.7 SqueezeNet network structure (source: arXiv:1602.07360)

3.2.3   *Performance*

The experimental results show that the size of the model decreases to less than 1/50 of AlexNet, but the accuracy does not decrease, and the accuracy of Top-1 improves to a certain extent, as given in Table 3.2.

3.3   Xception

The Xception (Chollet, 2016) network was released on the arXiv website in 2016 by the Google team.

3.3.1   *Core Idea*

Xcepetion's name means extreme inception Network, which is inspired by the Inception module of GoogleNet. The use of the term "limit" is due to the strong assumption that Xception assumes that the mapping of cross-channel correlation

Table 3.1 Parameters of each layer of the SqueezeNet network (source: arXiv:1602.07360)

Layer name/type	Output size	Filter size/stride (if not a fire layer)	Depth	$S_{1\times1}$ (#1 × 1 squeeze)	$e_{1\times1}$ (#1 × 1 expand)	$e_{3\times3}$ (#3 × 3 expand)	$S_{1\times1}$ sparsity	$e_{1\times1}$ sparsity	$e_{3\times3}$ sparsity	#bits	#parameter before pruning	#parameter after pruning
Input image	224 × 224 × 3										–	–
conv1	111 × 111 × 96	7 × 7/2(× 96)	1				100% (7 × 7)			6bit	14,208	14,208
maxpool1	55 × 55 × 96	3 × 3/2	0									
fire2	55 × 55 × 128		2	16	64	64	100%	100%	33%	6bit	11,920	5746
fire3	55 × 55 × 128		2	16	64	64	100%	100%	33%	6bit	12,432	6258
fire4	55 × 55 × 256		2	32	128	128	100%	100%	33%	6bit	45,344	20,646
maxpool4	27 × 27 × 256	3 × 3/2	0									
fire5	27 × 27 × 256		2	32	128	128	100%	100%	33%	6bit	49,440	24,742
fire6	27 × 27 × 384		2	48	192	192	100%	50%	33%	6bit	104,880	44,700
fire7	27 × 27 × 384		2	48	192	192	50%	100%	33%	6bit	111,024	46,236
fire8	27 × 27 × 512		2	64	256	256	100%	50%	33%	6bit	188,992	77,581
maxpool8	13 × 12 × 512	3 × 3/2	0									
fire9	13 × 13 × 512		2	64	256	256	50%	100%	30%	6bit	197,184	77,581
conv10	13 × 13 × 1000	1 × 1/1(×1000)	1				20% (3 × 3)			6bit	513,000	103,400
avgpool10	1 × 1 × 1000	13 × 13/1	0									
Activations	Parameters						Compression info				1,248,424 (total)	421,098 (total)

Table 3.2 Size and accuracy of the SqueezeNet network (source: arXiv:1602.07360)

CNN architecture	Compression approach	Date type	Original → Compressed model size	Reduction in model size vs. AlexNet	Top-1 ImageNet accuracy	Top-5 ImageNet accuracy
AlexNet	None (baseline)	32 bit	240 MB	1x	57.2%	80.30%
AlexNet	SVD	32 bit	240 MB → 48 MB	5x	56.00%	79.4%
AlexNet	Network Pruning	32 bit	240,548 → 27 MB	9x	57.2%	80.3%
AlexNet	Deep Compression	5–8 bit	240 MB → 6.9 MB	35x	57.20%	80.30%
SqueezeNet(ours)	None	32 bit	4.8 MB	50x	57.50%	80.3%
SqueezeNet(ours)	Deee Compression	8 bit	4.8 MB → 0.66 MB	363x	57.50%	80.30%
SqueezeNet(ours)	Deee Compression	6 bit	4.8 M8 → 0.47 MB	510x	57.50%	80.3%

and spatial correlation in the feature graph of a convolution neural network can be fully decoupled.

Xception is not a lightweight model in the true sense of the word, but it is the first to introduce the methods of depthwise convolution and pointwise convolution, collectively known as depthwise separable convolution, which has inspired many subsequent lightweight neural networks, which is worth referencing.

Xcepeion reduces the number of parameters, but it also broadens the network structure. Therefore, the model performs better than Inception v3 with the same number of parameters without compressing the model.

Xception is a series of extensions to inception v3.

First, the traditional Inception structure is simplified, as shown in Fig. 3.8.

The above structure can first use a unified 1 × 1 convolution kernel, and then the input of each 3 × 3 convolution kernel is only a part of the 1 × 1 convolution output feature map, i.e., 1/3, as shown in Fig. 3.9.

Based on the above, the 1 × 1 convolution output feature map is divided into channels, and each channel corresponds to a 3 × 3 convolution, that is, the number of 3 × 3 convolution kernels is the number of channels of the 1 × 1 convolution feature map, as shown in Fig. 3.10.

This extreme form of the Inception module is called the Xception module, and it is almost identical to the depthwise separable convolution module, but with two minor differences:

1. In the order of operations, the Xception module first performs convolution, and then performs depthwise convolution, while the depthwise separable convolution module does the opposite.
2. In Xception, there is a ReLU activation function after 1 × 1 convolution and 3 × 3 convolution, but there is no depthwise separable convolution.

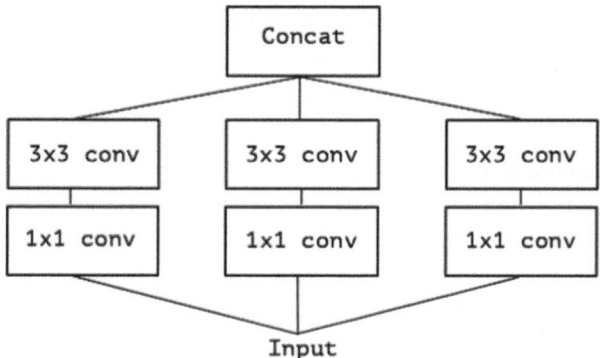

Fig. 3.8 Simplified Inception module (source: arXiv:1610.02357)

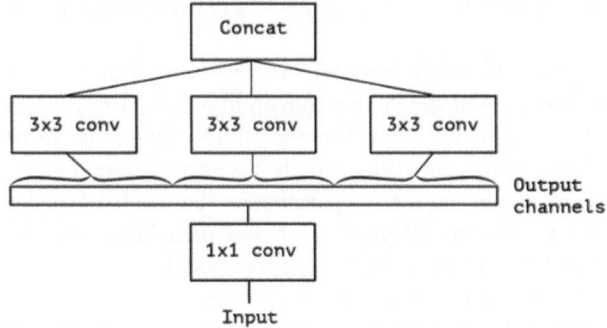

Fig. 3.9 Strict equivalence form of the Inception simplified module (source: arXiv:1610.02357)

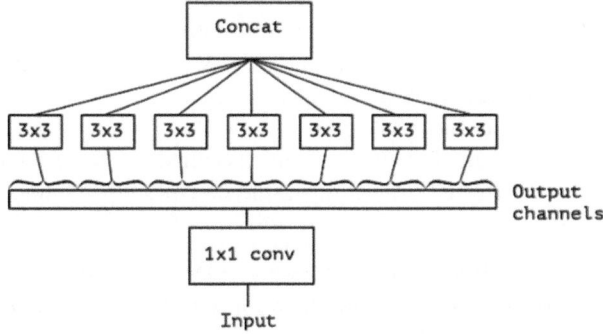

Fig. 3.10 Simplified the "extreme" form of the Inception module (source: arXiv:1610.02357)

3.3.2   Network Structure

The data first goes through the entry flow, then through the middle flow that is repeated eight times, and finally through the exit flow. Note that all convolution and depthwise separable convolution layers are then batch normalized. All depthwise separable convolution layers use a depth multiplier of 1 (no depth expansion), as shown in Fig. 3.11.

3.3.3   Performance

Compared with VGG-16, ResNet-152, Inception V3, and other networks, the accuracy of Xception network has been improved. The details are given in Table 3.3.

At the same time, the parameters of the Xception network are slightly reduced compared with the Inception network, and the training speed is also slightly improved, as given in Table 3.4.

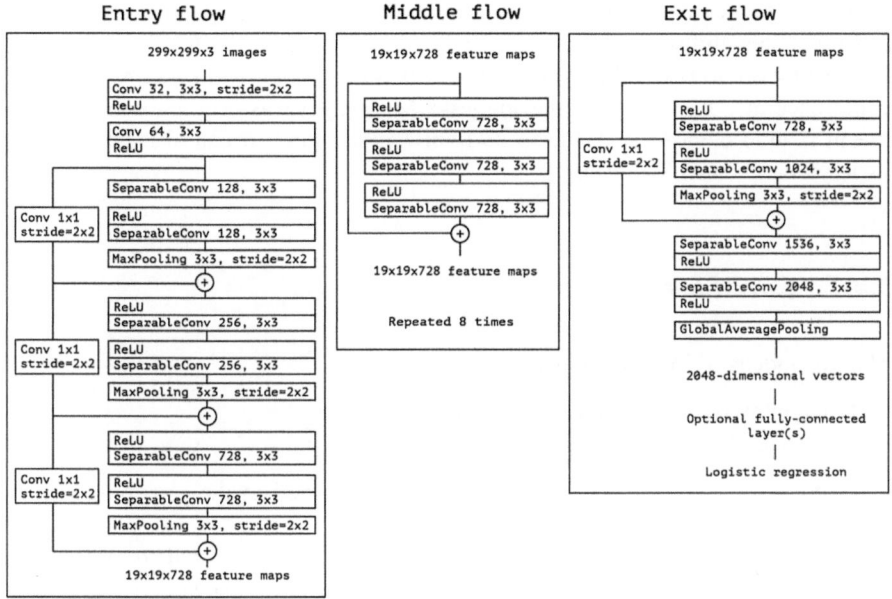

Fig. 3.11   Xception network structure (source: arXiv:1610.02357)

Table 3.3 Comparison of accuracy of Xception with other networks (source: arXiv:1610.02357)

	Top-1 accuracy	Top-5 accuracy
VGG-16	0.715	0.901
ResNet-152	0.770	0.933
Inception V3	0.782	0.941
Xception	**0.790**	**0.945**

Table 3.4 Number of parameters and training speed of Xception network (source: arXiv:1610.02357)

	Parameter count	Steps/second
Inception V3	23,626,728	31
Xception	**22,855,952**	**28**

3.4   MobileNet V1

MobileNet (Howard et al., 2017) is an efficient model proposed for embedded and mobile devices that aims to perform inference and improve accuracy using limited computing resources. It has the characteristics of small size, low latency, and low-power consumption. It can be used like large-scale models (e.g., Inception) for tasks such as classification, detection, semantic segmentation, etc.

Different from the principle of model compression, MobileNet directly trains small models, and it considers resource constraints, such as the latency and size of the model, to reconstruct the lightweight network from the perspective of depthwise separable convolution.

The first version of MobileNet was released by the Google team at the CVPR-2017 conference in 2017. As the name suggests, MobileNet was originally intended for use in mobile devices, and due to its lightweight nature, it can also be used in a variety of embedded devices.

3.4.1   Core Idea

The main ideas of MobileNet are:

1. Use depthwise separable convolution.
2. set the depth multiplier and resolution multiplier.

Depthwise separable convolution has been described in the previous chapters and will not be repeated.

To further compress the network, the number of channels and the size of the feature map can be further reduced. Reduce the number of channels by depth multiples and reduce size by resolution multiples.

Depth Multiplier α

The function of this multiple is to reduce the number of input and output channels in equal proportions, and the value is (0, 1]. Assuming that the original number of input channels is C, and the number of output channels is O, then this multiple changes the number of input channels to αC and the number of output channels to αO. After applying this multiplier,

The computational cost decreases to:

$$H \times W \times \alpha O \times K^2 + H \times W \times \alpha C \times \alpha O$$

The number of parameters is:

$$\alpha O \times K^2 + C \times \alpha O$$

The Resolution Multiple β

The function of this multiple is to reduce the resolution of the input image, and thus the resolution of the middle-layer feature map. The value in the example is (0, 1]. After applying this multiplier,

The computational cost decreases to:

$$\beta H \times \beta W \times O \times K^2 + \beta H \times \beta W \times C \times O$$

3.4.2   Network Structure

The MobileNet structure is built on depthwise separable convolution, which is implemented by the following basic building blocks.

Basic Building Blocks

The basic building blocks are shown in Fig. 3.12, and on the left side of the figure is a standard convolution layer, consisting of a 3×3 convolution, batch normalization, and the ReLU activation function. On the right side of the figure is a depthwise separable convolution, consisting of a 3×3 depthwise convolution (following batch normalization and ReLU activation) and a 1×1 pointwise convolution (also following batch normalization and ReLU activation).

Network Structure

Calculating depthwise convolution and pointwise convolution as separate layers, so that MobileNet has a total of 28 layers. The first layer is a complete convolution, after which there are multiple depthwise separable convolution layers, all layers follow a batch normalization and ReLU nonlinear activation, the average pooling layer reduces the spatial resolution to 1 before the fully connected layer, and there is no nonlinear activation layer after the fully connected layer, which is directly imported into the softmax layer for classification, as given in Table 3.5.

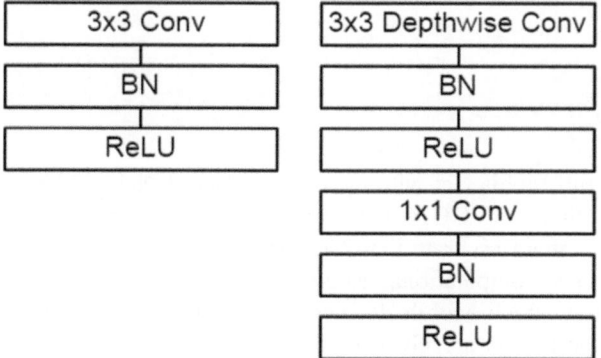

Fig. 3.12   Basic building blocks of MobileNet v1 (source: arXiv:1704.04861)

Table 3.5 MobileNet v1 network structure (source: arXiv:1704.04861)

Type/Stride		Filter Shape	Input Size
Conv/s2		$3 \times 3 \times 3 \times 32$	$224 \times 224 \times 3$
Conv dw/s1		$3 \times 3 \times 32dw$	$112 \times 112 \times 32$
Conv/s1		$1 \times 1 \times 32 \times 64$	$112 \times 112 \times 32$
Conv dw/s2		$3 \times 3 \times 64dw$	$112 \times 112 \times 64$
Conv/s1		$1 \times 1 \times 64 \times 128$	$56 \times 56 \times 64$
Conv dw/s1		$3 \times 3 \times 128dw$	$56 \times 56 \times 128$
Conv/s1		$1 \times 1 \times 128 \times 128$	$56 \times 56 \times 128$
Conv dw/s2		$3 \times 3 \times 128dw$	$56 \times 56 \times 128$
Conv/s1		$1 \times 1 \times 128 \times 256$	$28 \times 28 \times 128$
Conv dw/s1		$3 \times 3 \times 256dw$	$28 \times 28 \times 256$
Conv/s1		$1 \times 1 \times 256 \times 256$	$28 \times 28 \times 256$
Conv dw/s2		$3 \times 3 \times 256dw$	$28 \times 28 \times 256$
Conv/s1		$1 \times 1 \times 256 \times 512$	$14 \times 14 \times 256$
5×	Conv dw/s1	$3 \times 3 \times 512dw$	$14 \times 14 \times 512$
	Conv/s1	$1 \times 1 \times 512 \times 512$	$14 \times 14 \times 512$
Conv dw/s2		$3 \times 3 \times 512dw$	$14 \times 14 \times 512$
Conv/s1		$1 \times 1 \times 512 \times 1024$	$7 \times 7 \times 512$
Conv dw/s2		$3 \times 3 \times 1024dw$	$7 \times 7 \times 1024$
Conv/s1		$1 \times 1 \times 1024 \times 1024$	$7 \times 7 \times 1024$
Avg Pool/s1		Pool 7×7	$7 \times 7 \times 1024$
FC/s1		1024×1000	$1 \times 1 \times 1024$
Softmax/s1		Classifier	$1 \times 1 \times 1000$

Training Details

- Use the RMSprop optimizer.
- A large amount of data augmentation was not done because the number of parameters was small, and the overfitting was not serious.
- Random image cropping input is used.
- Use a smaller weight decay factor or don't use it.

3.4.3   Performance

Compare the full MobileNet with the original GoogleNet and VGG16. MobileNet is almost as accurate as VGG16, but 32 times smaller in size and 27 times less computational. It is more accurate than GoogleNet, but it is smaller in size and more than 2.5 times less computational, as given in Table 3.6.

The speed at which MobileNet performs an image inference is tested on an iPhone, and the results are given in Table 3.7 in seconds. NE stands for neural engine on mobile devices.

Table 3.6 Accuracy, computational cost, and parameters of MobileNet v1 (source: arXiv:1704.04861)

Model	ImageNet accuracy	Million Mult-Adds	Million parameters
1.0 MobileNet-224	70.6%	569	4.2
GoogleNet	69.8%	1550	6.8
VGG 16	71.5%	15,300	138

Table 3.7 Image inference performance of MobileNet v1

	CPU	GPU	NE
iPhone 11	0.0193	0.0243	0.0050
iPhone XS	0.0226	0.0362	0.0064
iPhone X	0.0237	0.0374	n/a

Due to some limitations of the GPU on mobile devices, and the optimization of the CPU by CoreML on the iPhone, the inference speed on the CPU is due to the GPU, but NE brings a significant performance improvement.

3.5 MobileNet V2

MobileNet v2 (Sandler et al., 2018) was released in 2018 by the Google team at the CVPR2018 conference.

3.5.1 Core Idea

MobileNet v2 inherits the depthwise separable convolution of v1, however, v2 introduces two new features to the architecture:

1. Linear Bottleneck.
2. Inverted Residual Connection.

Linear Bottlenecks

The linear bottleneck refers to the structure shown in Fig. 3.13:

The depthwise convolution is now in the middle. The depthwise convolution layer is preceded by a 1 × 1 convolution, known as the extended layer. This increases the number of channels. The depthwise convolution layer is followed by another 1 × 1 convolution, which again reduces the number of channels, called the bottleneck layer (which is the exact opposite of what SqueezeNet does: SqueezeNet shrinks first and then expands). Note that there is no activation function after the

Fig. 3.13 Composition of
linear bottlenecks

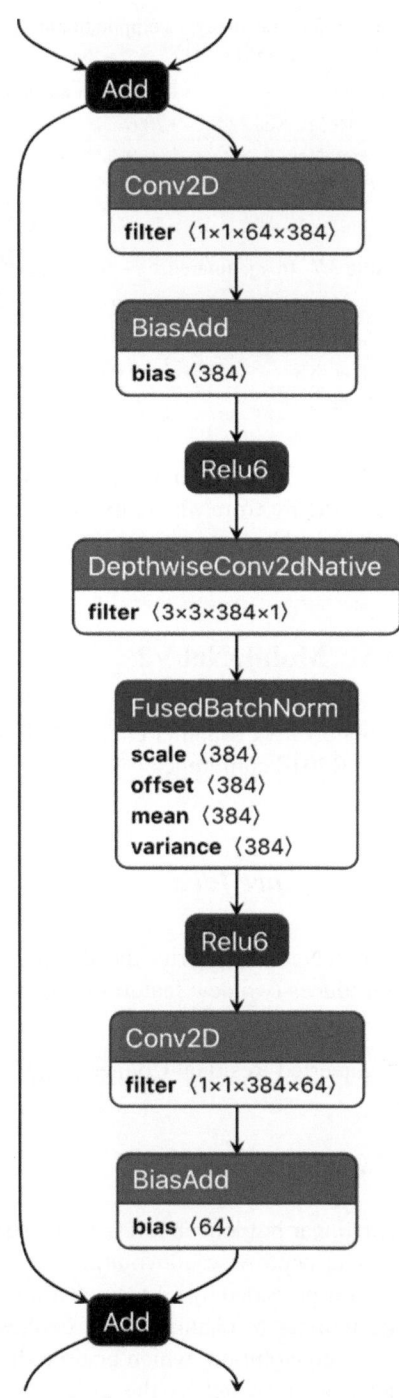

bottleneck layer, so it is called a linear bottleneck. Because the authors of MobileNet v2 found that if a nonlinear activation function (e.g., ReLU) is applied after the bottleneck layer, useful information is broken.

Inverted Residual Connection

The inverted residual connection exists between the bottleneck layers described above. Like traditional residual connections, inverted residual connections speed up training and improve accuracy. The authors of MobileNet v2 call it an inverted residual because it exists between the bottleneck layers, which have only a few channels. Conversely, the normal residual connection of ResNet exists between layers with many channels, as shown in Fig. 3.14.

3.5.2 Network Structure

The complete MobileNet v2 network structure consists of 17 consecutive bottleneck blocks, followed by regular 1 × 1 convolution, a global average pooling layer, and a classification layer, as given in Table 3.8.

(a) Residual block   (b) Inverted residual block

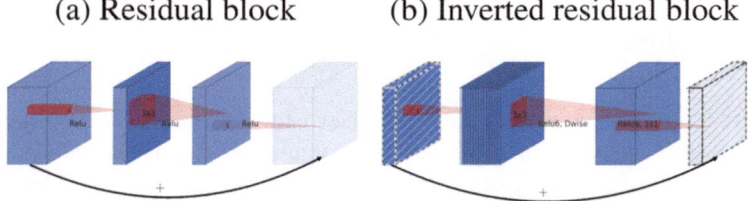

Fig. 3.14 Schematic diagram of the inverted residual connection (source: arXiv:1801.04381)

Table 3.8 MobileNet v2 network structure (source: arXiv:1801.04381)

Input	Operator	t	c	n	s
$224^2 \times 3$	conv2d	–	32	1	2
$112^2 \times 32$	bottleneck	1	16	1	1
$112^2 \times 16$	bottleneck	6	24	2	2
$56^2 \times 24$	bottleneck	6	32	3	2
$28^2 \times 32$	bottleneck	6	64	4	2
$14^2 \times 64$	bottleneck	6	96	3	1
$14^2 \times 96$	bottleneck	6	160	3	2
$7^2 \times 160$	bottleneck	6	320	1	1
$7^2 \times 320$	conv2d 1×1	–	1280	1	1
$7^2 \times 1280$	avgpool 7×7	–	–	1	–
$1 \times 1 \times 1280$	conv2d 1×1	–	k	–	

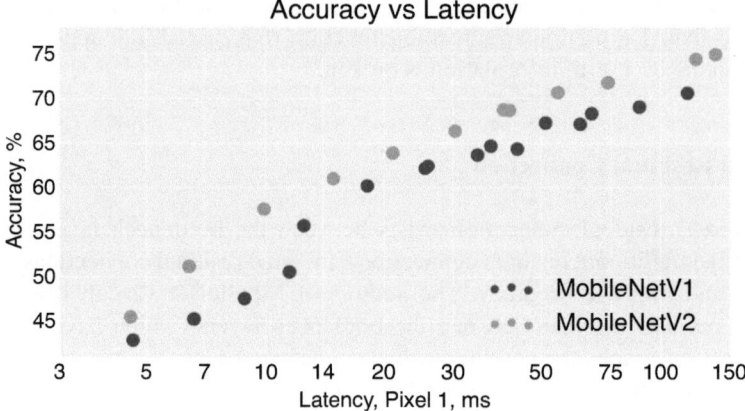

Fig. 3.15 Comparison of the performance of MobileNet v2 and v1

Table 3.9 Image inference performance of MobileNet v2

	CPU	GPU	NE
iPhone 11	0.0190	0.0185	0.0051
iPhone XS	0.0229	0.0284	0.0064
iPhone X	0.0276	0.0357	n/a

3.5.3   Performance

Compared with MobileNet v1, the MobileNet v2 model has faster inference speed with the same accuracy. In particular, the calculation amount is reduced by two times, the number of parameters is reduced by 30%, and the speed on Google Pixel phones is increased by about 30–40%, while also achieving greater accuracy, as shown in Fig. 3.15.

However, the experimental results on the iPhone are almost the same as those of MobileNet v1, as given in Table 3.9. This could be that v2 has more layers and is related to the specific implementation of the IOS CoreML library.

3.6   MnasNet

Before introducing MobileNet v3 (Tan et al., 2019), it is necessary to introduce MnasNet, which was released in 2019 by the Google team at the CVPR 2019 conference. Unlike the previous network, it was discovered through Neural Network Architecture Search (NAS).

3.6.1 Core Idea

Use the Neural Network Architecture Search (NAS) to find mobile-friendly network structures. The goal is to find a model that achieves a good compromise between accuracy and latency. During the search, the latency of the potential new architecture is measured directly by executing it on the phone. This is called platform-aware neural network architecture search.

3.6.2 Network Structure

The neural network architecture search found model is shown in Fig. 3.16.

This is the MnasNet-A1 architecture. It is made up of many different building blocks and can be repeated many times. Some of them use convolution layers with 3×3 convolution kernels, others with 5×5 convolution kernels. This structure may seem arbitrary, but that's the best model architecture found in the search.

The first few layers of the model are:

- Regular 3×3 convolution with 32 filters.
- 3×3 depthwise convolution.
- 1×1 convolution with 16 filters and linear activation.

This is the Conv3×3 and SepConv modules in the figure above.
The next building block is called MBConv6 and contains:

- 1×1 expansion layer increases the number of channels by 6.
- 3×3 depthwise convolution.
- 1×1 bottleneck layer to reduce channels.
- Residual connections from the previous bottleneck layer.

Other building blocks in MnasNet are variants of the above modules, such as the MBConv3 block, where the extension layer uses multiples of 3 instead of 6, and the depthwise convolution layer uses 5×5 instead of 3×3.

The above structure is very similar to the function of MobileNet v2. The new thing is the addition of squeeze and excitation (SE) module. Note that the SE module is part of a human-specified network structure search space, not a neural network architecture search.

The structure of the SE module is shown in Fig. 3.17.

It works as follows: First, there is a global average pooling operation (the mean layer in the figure) that reduces the spatial size of the feature map to 1×1. Assuming that the dimensions of the feature map are $H \times W \times C$, it is now just a vector with C elements. This is called "squeezing."

Next comes the "excitation"part: this is performed by two 1×1 convolution layers with the ReLU activation function between them. The first convolution layer significantly reduces the number of channels by a factor of 12 or 24 and the second

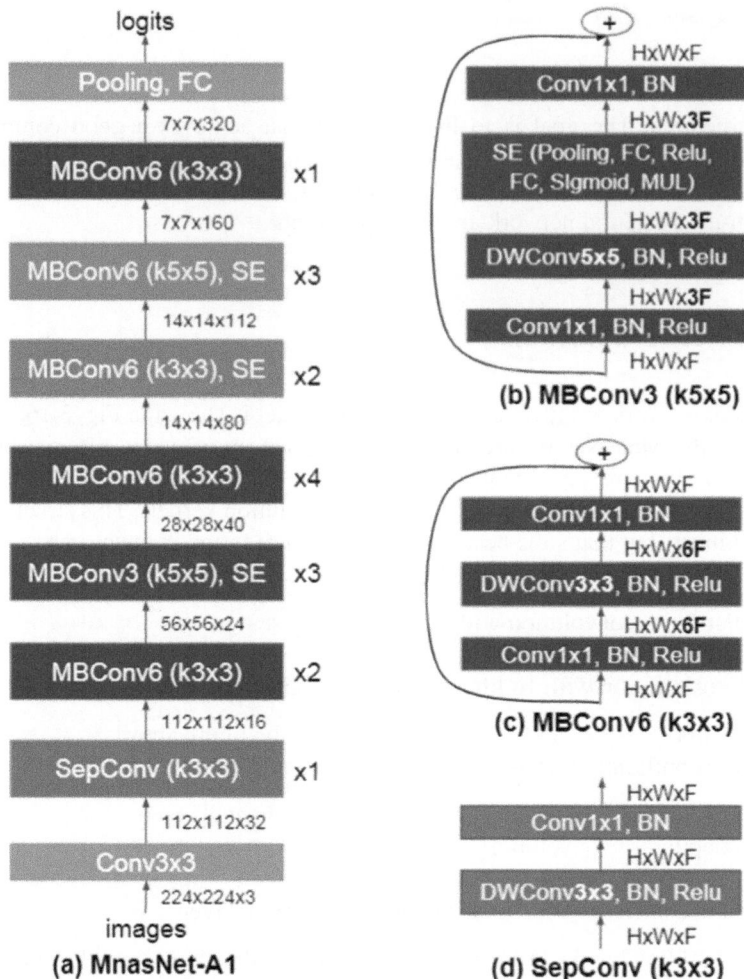

Fig. 3.16 MnasNet network structure

convolution layer increases the number of channels again to C. Finally, use the Sigmod function to scale the result between 0 and 1.

What is the purpose of all this? It is what the SE module will learn which features are important. It first compresses the channel vector and then tries to restore it, with the result that only the most important channels remain. As the original paper says: "... The SE module learns how to measure incoming feature maps."The SE module helps to improve the accuracy of the model and is very cheap because it is small.

The final search for the MnasNet architecture is essentially like MobileNet v2, except for details such as the size of the convolution filter and the SE structure. This is no coincidence, as the search space was heavily affected by MobileNet from the start.

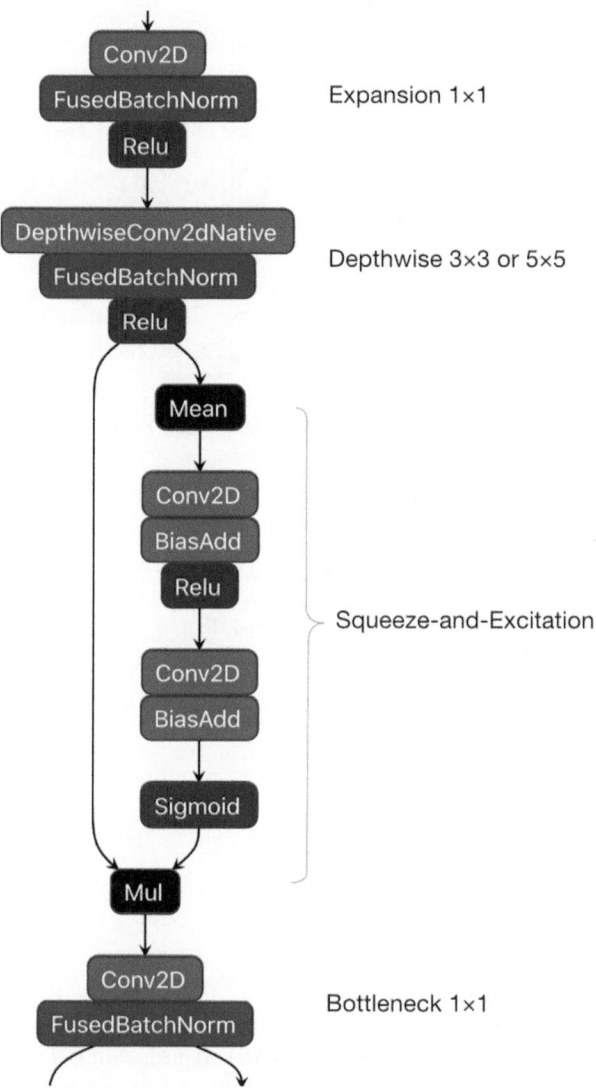

Fig. 3.17 Squeeze–excitation module

3.6.3   Performance

MnasNet can find models that ran 1.5 times faster than MobileNet v2 while achieving the same Top-1 accuracy, as shown in Fig. 3.18.

The image inference speed was tested on the iPhone, and the results are given in Table 3.10 in seconds.

Compared with MobileNet, there is a 10–20% speed increase.

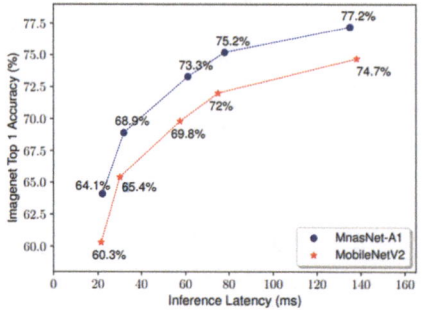 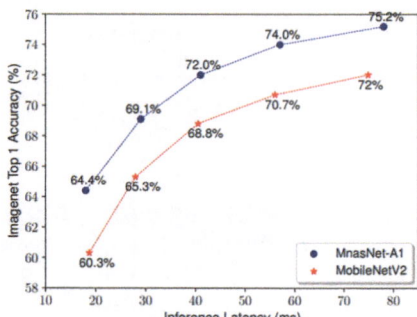

(a) Depth multiplier = 0.35, 0.5, 0.75, 1.0, 1.4, corresponding to points from left to right.

(b) Input size = 96, 128, 160, 192, 224, corresponding to points from left to right.

Fig. 3.18 Accuracy and inference latency of MnasNet (source: arXiv:1807.11626)

Table 3.10 Inference speed of MnasNet images

	CPU	GPU	NE
iPhone 11	0.0180	0.0166	0.0136
iPhone XS	0.0219	0.0225	0.0195
iPhone X	0.0241	0.0318	n/a

3.7 MobileNet V3

MobileNet (Howard et al., 2019) was released in 2019 by the Google team at the ICCV 2019 conference.

3.7.1 Core Idea

Unlike the previous two versions, MobileNet v3 is of a combination of manual and automated network architecture search (NAS). It uses MnasNet-A1 as a starting point, but is optimized using the NetAdapt algorithm, which automatically simplifies the pre-trained model until a given delay is reached, while maintaining a high accuracy rate.

At the same time, MobileNet v3 has made a lot of manual improvements to MnasNet. The main changes are:

- Redesigned some expensive layers in MnasNet.
- Use Swish activation function instead of ReLU6.

Both MobileNet v1 and v2 start with a regular 3 × 3 convolution layer with 32 filters. As it turns out, this is a relatively slow layer. It has only a small amount of weight, but needs to work on large 224 × 224 feature maps. Experiments have shown that the use of 16 filters is sufficient. This reduces the number of parameters and increases the speed.

In earlier versions of MobileNet used ReLU6 as the activation function, while v3 used a specific version of the Swish activation function called hard swish or h-swish:

$$h\_swish(x) = x^* \, ReLU6(x \, 3 \, / \, 6$$

The creators of MobileNet discovered that h-swish was only useful on a deeper level. Larger in the feature map in shallow layers, so the computation cost of the activation function is higher, and simply using ReLU on these layers is cheaper than h-swish.

3.7.2 Network Structure

As mentioned above, MobileNet v3 uses the same structure as MnasNet-A1. The main building blocks are shown in Fig. 3.19.

This building block has a few minor differences from MnasNet:

- Activation is h-swish (the shallow layer is ReLU).
- The number of filters used by the extension layer is different (the optimal value of these filters is found using the NetAdapt algorithm).
- The number of channels output by the bottleneck layer may be different (also found by NetAdapt).
- The Squeeze and Excitation (SE) module only reduces the number of channels by 3 or 4 times.
- The SE module uses the ReLU6(x + 3)/6 formula as a rough approximation of the Sigmod function.

In MobileNet v2, before the global average pooling layer, there is a 1 × 1 convolution that scales the number of channels from 320 to 1280, so the classifier layer can use a lot of features, but it is also relatively slow.

In MobileNet v3, this layer sits behind the global average pooling layer, so it is faster to use a smaller feature map (1 × 1 instead of 7 × 7). This small change also allows us to remove the previous bottleneck layer and the depthwise 1 convolution layer. In total, it is possible to remove three expensive layers without loss of accuracy.

The model output phase is shown in Fig. 3.20.

There are several variants of MobileNet v3 such as:

- Large: as described above.
- Small: fewer building blocks and fewer filters.
- Large Minimalist: Like "large", but without SE, h-swish, 5 × 5 convolution.
- Small Minimalist: Like "small", but without SE, h-swish, 5 × 5 convolution.

The parameters of each layer of the large model are given in Table 3.11, and the rest of the variants are not repeated.

Fig. 3.19 Building blocks
of MobileNet v3

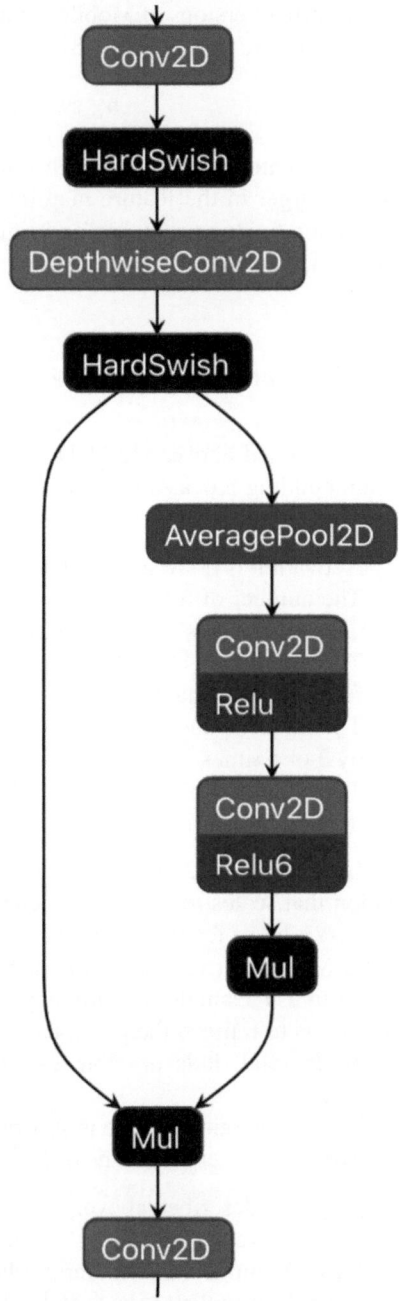

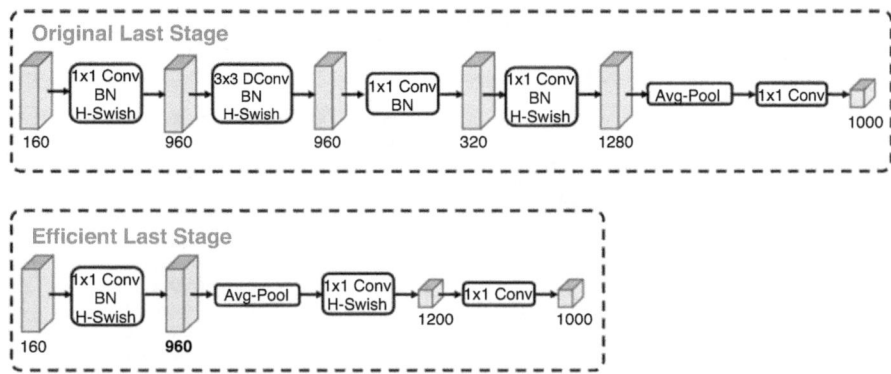

Fig. 3.20   MobileNet v3 output stage structure optimization (source: arXiv:1905.02244)

Table 3.11   MobileNet v3 model structure (source: arXiv:1905.02244)

Input	Operator	Exp size	#out	SE	NL	s
$224^2 \times 3$	conv2d	-	16	–	HS	2
$112^2 \times 16$	bneck, 3×3	16	16	–	RE	1
$112^2 \times 16$	bneck, 3×3	64	24	-	RE	2
$56^2 \times 24$	bneck, 3×3	72	24	–	RE	1
$56^2 \times 24$	bneck, 5×5	72	40	✓	RE	2
$28^2 \times 40$	bneck, 5×5	120	40	✓	RE	1
$28^2 \times 40$	bneck, 5×5	120	40	✓	RE	1
$28^2 \times 40$	bneck, 3×3	240	80	–	HS	2
$14^2 \times 80$	bneck, 3×3	200	80	–	HS	1
$14^2 \times 80$	bneck, 3×3	184	80	–	HS	1
$14^2 \times 80$	bneck, 3×3	184	80	–	HS	1
$14^2 \times 80$	bneck, 3×3	480	112	✓	HS	1
$14^2 \times 112$	bneck, 3×3	672	112	✓	HS	1
$14^2 \times 112$	bneck, 5×5	672	160	✓	HS	2
$7^2 \times 160$	bneck, 5×5	960	160	✓	HS	1
$7^2 \times 160$	bneck, 5×5	960	160	✓	HS	1
$7^2 \times 160$	conv2d, $I \times 1$	–	960	–	HS	1
$7^2 \times 960$	pool, 7×7	–	–	–	–	1
$1^2 \times 960$	conv2d 1×1, NBN	–	1280	–	HS	1
$1^2 \times 1280$	conv2d 1×1, NBN	–	k	–	–	1

3.7.3   *Performance*

Thanks to these additional improvements, MobileNet v3 is still faster than MnasNet and has the same accuracy even though MobileNet v3 uses the same type of building blocks and has more parameters than MnasNet!

Figure 3.21 compares the performance of MobileNet v3 with similar models.

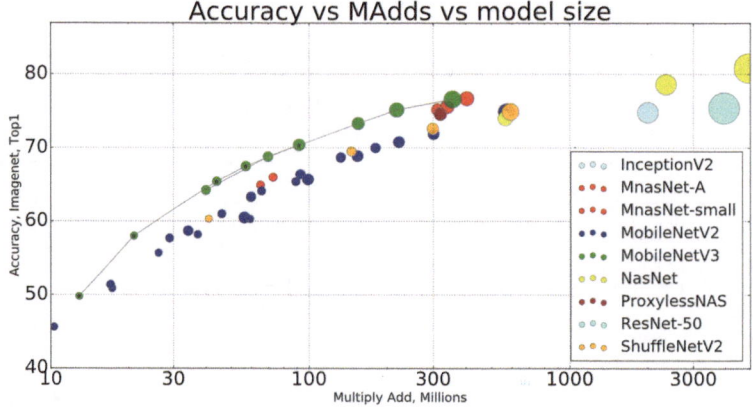

Fig. 3.21 MobileNet v3 performance comparison (source: arXiv:1905.02244)

Table 3.12 MobileNet v3 image inference speed

		CPU	GPU	NE
MobileNet v3 (Small)	iPhone 11	0.0063	0.0106	0.0102
	iPhone XS	0.0071	0.0176	0.0164
	iPhone X	0.0093	0.0211	Not applicable
MobileNet v3 (Large)	iPhone 11	0.0152	0.0192	0.0169
	iPhone XS	0.0174	0.0264	0.0221
	iPhone X	0.0219	0.0332	Not applicable

For the ImageNet classification task, MobileNet v3-Large delivers 3.2% higher accuracy while reducing latency by 20% compared to MobileNet v2. Compared to the MobileNet v2 model with similar latency, the accuracy of MobileNet v3-Small is improved by 6.6%.

For the COCO object detection task, MobileNetV3-Large has a detection speed of more than 25% faster than MobileNet v2.

For the Cityscapes target segmentation task, MobileNet v3-Large is 34% faster than MobileNet v2 with the same accuracy.

The image inference speed of MicroNet v3 was tested on iPhone, and the results are given in Table 3.12, in seconds.

Taking iPhone 11 as an example, v3-Small has about 3 times the CPU performance and more than 2 times the GPU performance compared to v1 and v2.

3.8   YOLO

YOLO (Redmon et al., 2016) (You Only Look Once) is a target detection algorithm proposed by Joseph Redmon et al. in 2016.

3.8.1   Core Idea

The core idea of YOLO is to transform the target detection task into a regression problem, and simultaneously locate and classify targets through a single neural network to achieve real-time and efficient target detection.

The principle of YOLO algorithm is as follows:

Meshing: First, the input image is divided into fixed-size grids. Each grid is responsible for detecting objects in that grid. Different sized images may result in different numbers of grids.

Bounding box prediction: Each grid predicts multiple bounding boxes (bounding boxes), each bounding box contains an object. A bounding box is represented by a set of coordinate values, including the center coordinate, width, and height of the bounding box.

Object classification: For each bounding box, use a classifier to predict the class of the object. Convolutional neural networks (CNNs) are usually used to extract features and fully connected layers are used for classification.

Confidence evaluation: Each bounding box also predicts a confidence score, which represents the probability that an object exists in that bounding box and the accuracy of the bounding box.

Non-maximum Suppression: For each category, use non-maximum suppression (NMS) to remove overlapping bounding boxes. The bounding box with the highest confidence is selected, and bounding boxes with high overlap with it are eliminated.

The advantage of the YOLO algorithm is its speed and real-time performance. Since YOLO transforms the target detection task into a single forward pass regression problem and does not require the generation process of region proposals, it can achieve real-time target detection. In addition, YOLO performs global optimization on the entire image and can capture the global context information of the object.

3.8.2   Network Structure

The structure of the Yolo network is shown in Fig. 3.22:

The input image size is 448*448. After several convolution layers and pooling layers, it becomes a 7*7*1024 tensor. Finally, after two layers of fully connected layers, the output tensor dimension is 7*7*30, which is the same as the general. There are not many differences between the convolutional object classification networks. The biggest difference is that the last fully connected layer of the classification network is generally connected to a one-dimensional vector. Different bits of the vector represent different categories, and the output vector here is a three-dimensional tensor. Quantity (7*7*30). This network structure is inspired by GoogLeNet. It does not use the BN layer and uses a layer of Dropout. Except for the

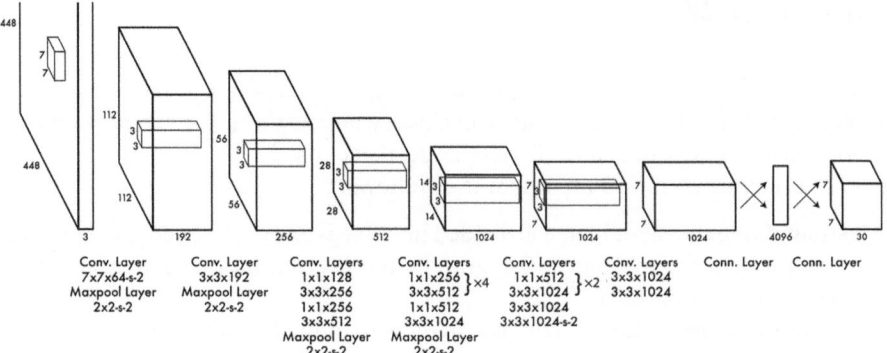

Fig. 3.22 Yolo network structure (Source: arXiv:1506.02640)

Table 3.13 Yolo-Fastest performance data (source: https://github.com/dog-qiuqiu/Yolo-Fastest)

Network	Model size	mAP(VOC 2007)	FLOPS
Tiny YOLOv2	60.5 MB	57.1%	6.97B
Tiny YOLOv3	33.4 MB	58.4%	5.52B
YOLO Nano	4.0 MB	69.1%	4.51B
MobileNetv2-SSD-Lite	13.8 MB	68.6%	~3B
MobileNetV2-YOLOv3	11.52 MB	70.20%	2.02B
Pelee-SSD	21.68 MB	70.09%	2.40B
Yolo Fastest	1.3 MB	61.02%	0.23B
Yolo Fastest-XL	3.5 MB	69.43%	0.70B
MobileNetv2-Yolo-Lite	8.0 MB	73.26%	1.80B

output of the last layer, which uses a linear activation function, all other layers use Leaky Relu activation functions.

3.8.3 Performance

The YOLO model has been improved many times and still maintains detection accuracy while continuously reducing the model size. Smaller size means faster inference speed to meet the needs of real-time video processing. Currently, the fastest YOLO algorithm—only 1.4 MB in size and 148 frames per second on a single CPU core. It is particularly easy to deploy on some mobile devices. The specific test results are as follows in Table 3.13.

Compared with MobileNet v2-Yolo v3, the number of parameters is only 1/10, but the mAP dropped by 7.5%. In other words, Yolo achieves lightweight models by sacrificing accuracy.

3.9 Applications of Lightweight Neural Networks

The above only introduces some of the most typical lightweight neural networks. In fact, there are many excellent lightweight neural networks, including:

- BlazeFace.
- SqueezeNext.
- ShuffleNet.
- CondenseNet.
- ESPNet.
- DiCENet.
- FBNet & ChamNet.
- GhostNet.
- MixNet.
- EfficientNet.

etc.

Of course, most of the models described above are image classifiers. However, there aren't many applications that just require a classifier. We want to use these models as feature extractors, as part of a large neural network that performs some kind of computer vision task. These tasks include object detection, pose estimation, semantic segmentation, and more.

In most cases, transfer learning is required when building new neural networks. You can use the above model as the backbone and pre-train it with a generic dataset such as ImageNet or COCO, so that you get an excellent feature extractor. Then, add some new layers on top and fine-tune these new layers with your own data. The result is a lightweight neural network that can accomplish the specific tasks you need.

In addition, in lightweight neural networks, the accuracy and speed of inference are often not compatible. Improving accuracy often requires sacrificing speed, and vice versa. When applying, you need to make trade-offs based on specific problems and select the most appropriate network structure.

In the process of the vigorous development of lightweight neural network technology, reducing the complexity of algorithms is an eternal theme, and new and better lightweight algorithms are about to emerge. In any case, these lightweight algorithms have the following benefits:

- Training speed is fast. And in distributed training, communication requirements with the server are minimal.
- Few parameters and small model size. Small storage space and fast download speeds.
- The inference speed is fast, and the requirements for chip computing power are low.

These features make lightweight neural networks ideal for deployment on embedded and mobile devices.

References

Chollet, F. (2016). Xception: deep Learning with depthwise separable convolutions. arXiv:1610.02357.

Howard, A. G., Zhu, M., Chen, B., Kalenichenko, D., Wang, W., Weyand, T., ... Adam, H. (2017). MobileNets: efficient convolutional neural networks for mobile vision applications. arXiv:1704.04861.

Howard, A., Sandler, M., Chu, G., Chen, L.-C., Chen, B., Tan, M., ... Adam, H. (2019). Searching for MobileNetV3. arXiv:1905.02244.

Iandola, F. N., Han, S., Moskewicz, M. W., Ashraf, K., Dally, W. J., & Keutzer, K. (2016). SqueezeNet: AlexNet-level accuracy with 50x fewer parameters and <0.5MB model size. arXiv:1602.07360.

Krizhevsky, A., Sutskever, I., & Hinton, G. E. (2017). ImageNet classification with deep convolutional. *Communications of the ACM, 60*(6), 84–90.

Redmon, J., Divvala, S., Girshick, R., & Farhadi, A. (2016). You only look once: unified, real-time object detection. arXiv:1506.02640.

Sandler, M., Howard, A., Zhu, M., Zhmoginov, A., & Chen, L.-C. (2018). MobileNetV2: inverted residuals and linear bottlenecks. Paper presented at the The IEEE Conference on Computer Vision and Pattern Recognition (CVPR).

Tan, M., Chen, B., Pang, R., Vasudevan, V., Sandler, M., Howard, A., & Le, Q. V. (2019). MnasNet: platform-aware neural architecture search for mobile. Paper presented at the The IEEE Conference on Computer Vision and Pattern Recognition (CVPR).

Chapter 4
Compression of Deep Neural Network

Abstract This chapter introduces the third and fourth components to implement embedded AI, model compression, and model compilation. Model compression is a method to reduce the size of deep neural networks without changing the network structure. Assuming that the neural network model has been generated, techniques such as pruning, weight sharing, quantization, binary/ternary, Winograd convolution, etc. can be used to "compress" the neural network. Model distillation is also introduced, a method that learns a small student model from a large teacher model but maintains the accuracy of the teacher model. Model compilation is the process of translating AI models into AI chip instructions. There are many optimization methods. This chapter focuses on the compression compilation co-design method. This method can effectively optimize the size and speed of deep neural network models and greatly shorten the adjustment time of the compression process, thereby enabling the deployment of deep neural network models on embedded devices.

Keywords Model compression · Model compilation · Pruning · Weight sharing · Quantization · Model distillation

4.1 General Approaches of Neural Network Compression

Today, deep neural network models require large amounts of computation, memory, and energy, which become a bottleneck in situations where we need real-time inference or run models on edge devices and browsers with limited computing resources. Energy efficiency is the focus of current deep neural network models. One way to solve this efficiency is to reduce the size of the neural network.

Larger neural network models require more memory references, which in turn require more energy. This section explores ways to reduce the size of deep neural networks without changing the network structure. That is, it is assumed that the neural network model has been generated, but some techniques are used to "compress" the neural network.

Specifically, you can start from the following directions (Han et al., 2015):

1. Pruning.
2. Weight sharing.
3. Quantization.
4. Binary/Ternary Net.
5. Winograd Convolution.

4.1.1   *Pruning*

Pruning is a method to accelerate neural network inference, which can effectively generate models that are smaller in size, more memory efficient, and more energy efficient, and the inference speed is faster with minimal loss of accuracy.

Deep learning draws inspiration from the field of neuroscience, and pruning in deep learning is also a biologically inspired concept. When a human baby is born, there are about 50 trillion synapses, and by the age of one, it has developed to 1000 trillion synapses, but by the time the basic intellectual development is completed, the number of synapses has dropped to 500 trillion. That is, some synapses are useless, and their elimination does not affect intelligence. This process occurs in all mammals, and humans only complete this process around the age of 20, as shown in Fig. 4.1.

Pruning of neural networks attempts to simulate this process, pruning means deleting the connection if the weight is below the threshold, and deleting a neuron if all the connections are deleted, then it is also deleted.

As shown in Fig. 4.2, a neural network typically looks like the one on the left. Each neuron between two adjacent layers is connected to each other, and this full connection means many floating-point multiplication operations. Ideally, if each neuron is connected to only a few other neurons, a lot of multiplication can be saved. This is called a "sparse" network. Sparse models are easier to compress, and

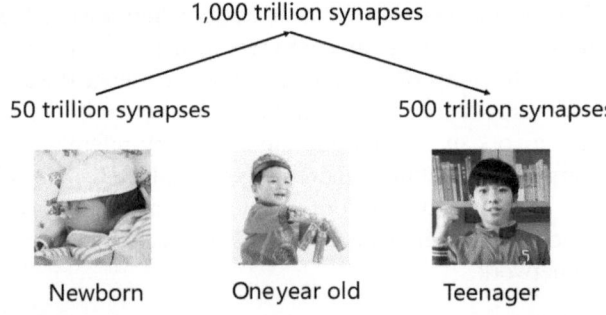

Fig. 4.1 Pruning of synapses in the human brain

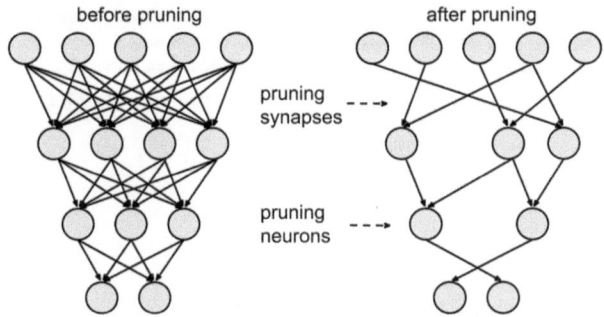

Fig. 4.2 Neural network pruning

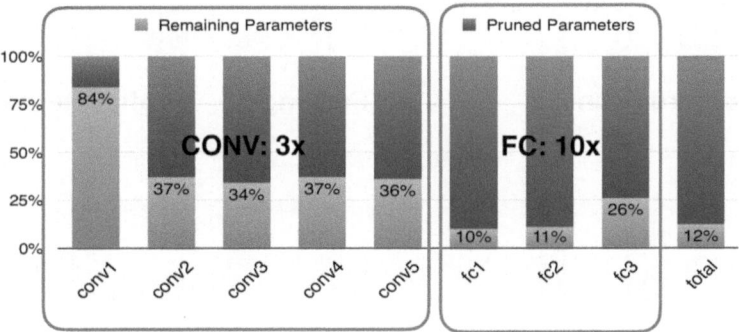

Fig. 4.3 AlexNet pruning efficiency

we can skip those deleted "zero" connections during inference, reducing the latency of the inference process.

Specifically, neurons are ranked according to their contribution to the neural network, and then the lower-ranked neurons are removed from the network, making the network smaller and faster. For example, they can be ranked based on the L1/L2 norm of the neuronal weights.

After pruning, the accuracy of the neural network will generally decrease, and the network will be restored by performing multiple training-pruning-training-pruning processes. If you prune too much at once, the network may be damaged too much to recover. Therefore, in practice, it is an iterative process, often called "Iterative pruning."

Taking the AlexNet model as an example, through the pruning algorithm, the connections of the convolutional layer are reduced to 1/3 at most, and the fully connected layer is reduced to 1/10 at most. Overall, the connection is reduced to 12% of the original, as shown in Fig. 4.3.

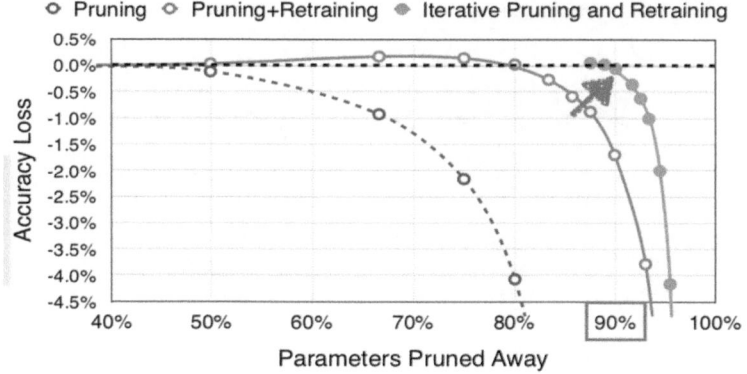

Fig. 4.4 Retraining after pruning to reduce accuracy loss

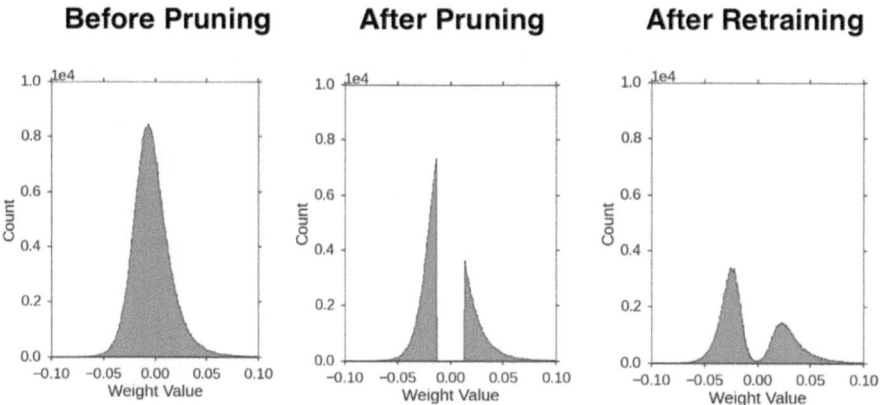

Fig. 4.5 Distribution of weight values before and after pruning

Of course, after pruning, the model needs to be retrained to restore accuracy, as shown in Fig. 4.4. After multiple iterations, more than 85% of the connections are pruned, and the loss of accuracy approaches 0.

The benefits of pruning are significant, and in the case of the fully connected layer, pruning can greatly improve the inference speed and energy efficiency. Inference speed is increased by three times, and energy efficiency is increased by six times.

Observing the distribution of weight values before and after pruning, it is possible to understand why such a large increase occurred, as shown in Fig. 4.5.

Before pruning, there are many weights approaching 0, and they do not contribute much to the accuracy of inference. These weights are removed directly after pruning, but the accuracy is reduced. With retraining, the number of weights has not changed, but their value distribution has become more average and smoother.

4.1.2 Weight Sharing

The principle of weight sharing is to cluster weights with approximate values together and store them in a dictionary, thereby reducing the storage space occupied by weights.

One of the simplest implementations is to round up the weights to 256 levels. With this approach, it is possible to reduce the size of a model from 87 MB to 26 MB with an accuracy reduction of only 1%.

The complete method is shown in Fig. 4.6. The four weights of 2.09, 2.12, 1.92, and 1.87 are close to each other, their centroids are 2, and they are clustered as the same weight 2, stored in the dictionary. Similarly, other weights are clustered to obtain a weight dictionary, which is a floating-point sequence matrix that can be multiplied by an integer matrix to approximate the original weight matrix. Compared to the original 4 × 4 floating-point matrix, it takes up less storage space.

At the same time, the gradient matrix also carries out the corresponding grouping operation, and the gradients corresponding to the above four weights are divided into a group, and then summed to obtain a new gradient column matrix, which is combined with the previous weight dictionary to obtain the optimized dictionary matrix.

Let's take a look at the distribution of weight values after weight sharing, as shown in Fig. 4.7.

On the left is the weight distribution after pruning, and on the right is the distribution after weight sharing. The weights with similar values are clustered into the

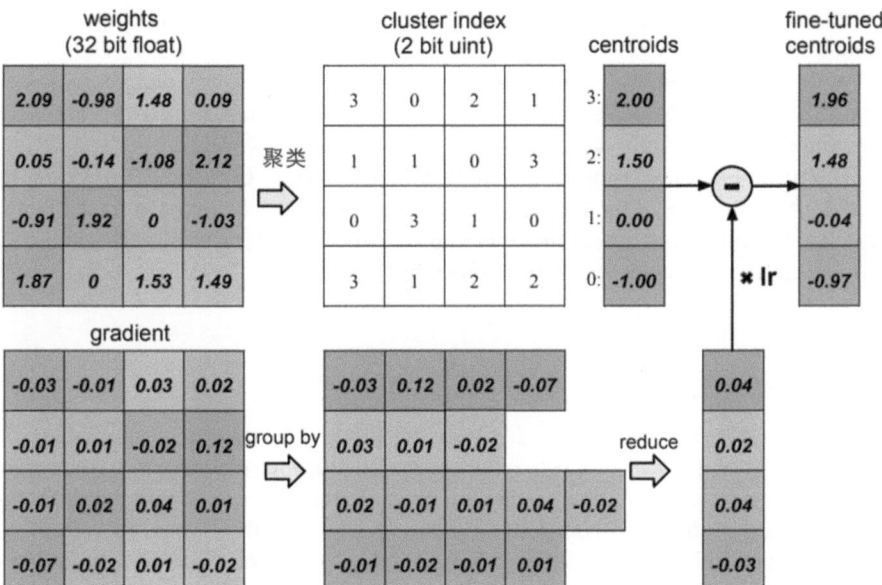

Fig. 4.6 Weight sharing

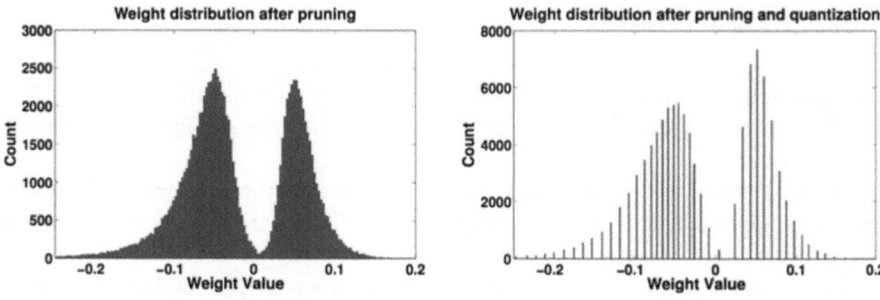

Fig. 4.7 Weight value distribution after weight sharing

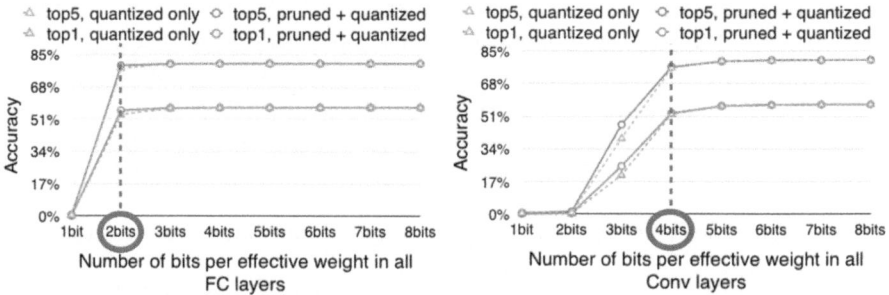

Fig. 4.8 The number of bits per weight decreases

same weight, and the original continuous weight value distribution becomes a discrete distribution.

With the above methods, the number of bits occupied by each weight decreases from 32 bits to 2 bits for the fully connected layer and from 32 bits to 4 bits for the convolutional layer without much loss of accuracy, as shown in Fig. 4.8.

The weight sharing algorithm is applicable to a variety of models. For example, for AlexNet, the size of the model is compressed from 240 M bytes to 6.9 M bytes, which is 35 times, and the accuracy is improved from 80.27% to 80.30%. The situation is similar for other models, as shown in Fig. 4.9.

4.1.3 Quantization

The principle of quantization is to quantize the floating-point numbers in the trained model into fixed-point numbers, and then perform neural network inference. This is because fixed-point arithmetic is much more efficient than floating-point arithmetic.

Both the weight value and the activation value are quantized. The specific process is shown in Fig. 4.10.

Network	Original Size	Compressed Size	Compression Ratio	Original Accuracy	Compressed Accuracy
LeNet-300	1070KB ⟶ 27KB		40x	98.36% ⟶ 98.42%	
LeNet-5	1720KB ⟶ 44KB		39x	99.20% ⟶ 99.26%	
AlexNet	240MB ⟶ 6.9MB		35x	80.27% ⟶ 80.30%	
VGGNet	550MB ⟶ 11.3MB		49x	88.68% ⟶ 89.09%	
GoogleNet	28MB ⟶ 2.8MB		10x	88.90% ⟶ 88.92%	
SqueezeNet	4.8MB ⟶ 0.47MB		10x	80.32% ⟶ 80.35%	

Fig. 4.9   Effect of weight sharing for various models

- Train the model in floating-point format.
- Quantify the weight and activation values in the trained model: First, collect statistics about the weight and activation values, and then select the appropriate decimal point position.
- Fine-tune the model in floating-point format.
- Convert both the weight value and the activation value to a fixed-point format, which can be 32-bit, 16-bit, or 8-bit fixed-point number.

Because this method only ignores the remainder after a decimal point, the accuracy of the quantized model only decreases slightly, as shown in Fig. 4.11, using the quantization method on GoogleNet and VGG-16 networks, respectively, it can be seen that the accuracy does not decrease when quantified at a 32-bit fixed point, and the accuracy decreases slightly when quantified with a 16-bit fixed point, but it is acceptable.

4.1.4   Binary/Ternary

The principle of binary/ternary is to simplify the weights to two values of $-1, +1$ or three values of $-1, 0, +1$, as shown in Fig. 4.12. This will greatly speed up the inference of the neural network, because the convolution operation only needs to be approximated by addition and subtraction.

Specifically:

- Both the weight values and input values are binarized or ternarized.
- Convolution is implemented by binary dot product operations, with only addition and subtraction.
- Furthermore, the addition and subtraction of dot product operations can also be implemented through XOR and bit counting operations, to further improve the operation efficiency.

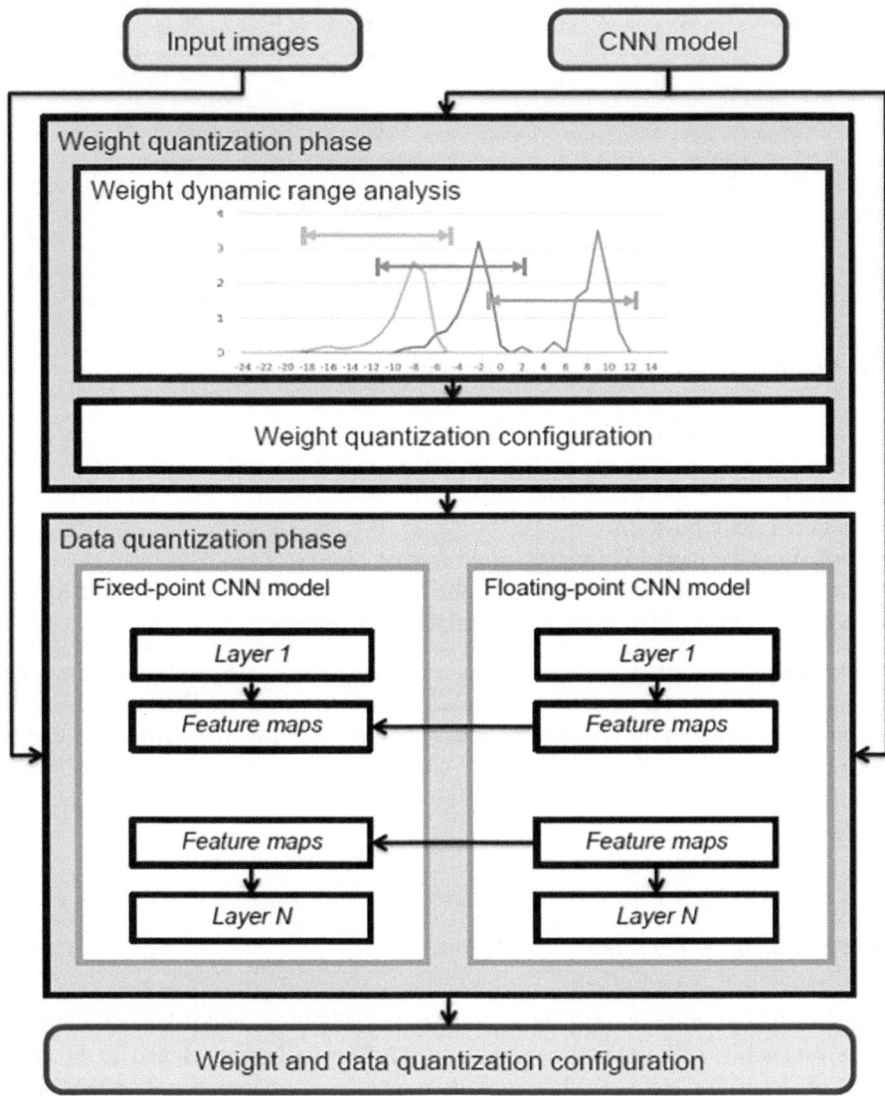

Fig. 4.10 Quantization process

By binarizing the weights, the memory occupied during inference is reduced by 32 times, and the computation speed is increased by two times, without any loss of accuracy. If the addition and subtraction operations in the convolution are further replaced by XOR-bit counting, the computation speed can be further increased to 58 times, and of course, the accuracy can be reduced from 56.7% to 44.2%.

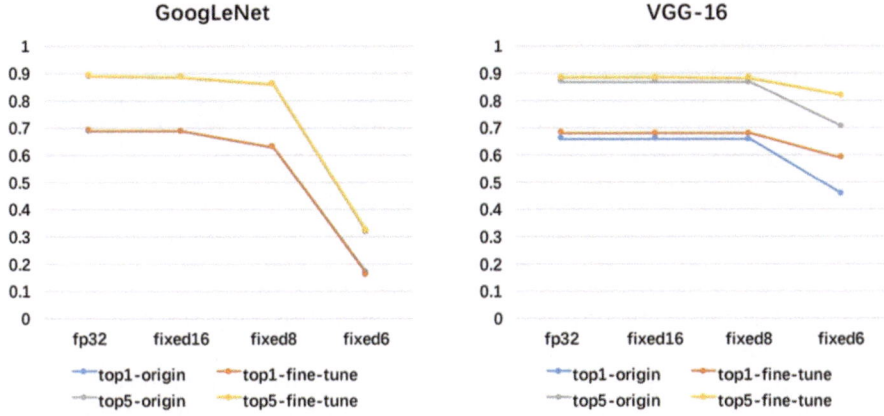

Fig. 4.11   Change in the accuracy of the model after quantization

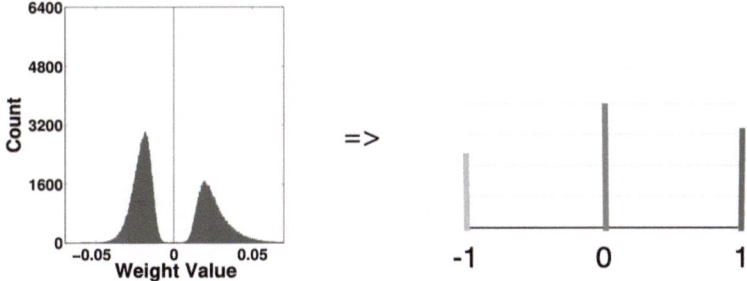

Fig. 4.12   Three-valued weight representation

4.1.5   Winograd Convolution

The principle of Winograd convolution (or Winograd transformation) is to use the Winograd algorithm to replace the time-consuming multiplication operation in the convolution operation with a less time-consuming addition and subtraction operation, to achieve the purpose of reducing algorithm time complexity.

The Winograd algorithm was first proposed by Shmuel Winograd in 1980 in his paper "Arithmetic complexity of computations," which is mainly used to reduce the computation of FIR filters.

The algorithm is like FFT (Fast Fourier Transform) in that it maps data to another space (unlike FFT, the Winograd algorithm maps data to a real space instead of a complex space). Addition and subtraction operations are used instead of partial multiplication operations. Since the addition and subtraction operations are much faster than the multiplication operations, a significant acceleration effect is achieved.

For example, to directly implement a FIR filter F(m,r) with m outputs and r parameters, a total of m × r times multiplication is required. However, with the

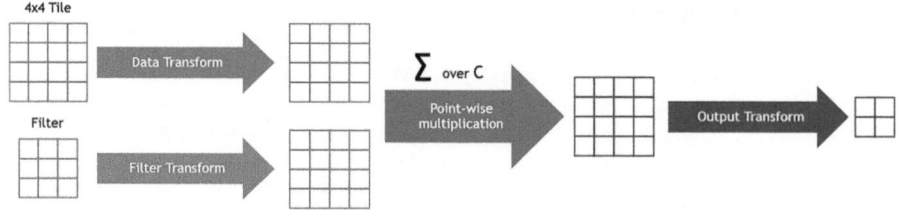

Fig. 4.13 Winograd convolution

Winograd algorithm, ignoring the transformation process, only $m + r - 1$ multiplication is required.

The method of applying the Winograd algorithm to the convolution operation is shown in Fig. 4.13, the input feature map and the convolution kernel are both transformed by Winograd, and then the pointwise multiplication operation is performed, and the dot product results of multiple channels are added, and then the inverse Winograd transformation of is performed to obtain the final output.

For example, if you use direct convolution: four outputs require $9 \times C \times 4 = 36 \times C$ multiplication and addition operations, where C is the number of channels. With Winograd convolution, only $16 \times C$ times of multiplication and addition are required for four outputs, which reduces the computational effort by 2.25 times.

Winograd convolutions are usually implemented by hardware.

4.1.6 Model Distillation

Hinton proposed the concept of knowledge distillation at NIPS2014 (Hinton et al., 2015), which aims to transfer the knowledge learned from a large model or multiple model ensembles to another lightweight single model to facilitate deployment. Simply put, it is to use a new small model to learn the prediction results of the large model.

Introduce the following nouns:

Teacher model—original model or model ensemble.
Student model—new model, also called **distilled model**, generally smaller than teacher model.

The core idea of distillation is that a good model is not designed to fit training data but to learn how to generalize to new data. The goal of distillation is to let students learn the generalization ability of the teacher. In theory, the results obtained will be better than those of students who simply fit the training data, because the teacher has obtained generalization ability from the training data. Taking the classification task as an example, the entropy of the teacher's prediction is higher than the entropy of the sample labels. Obviously, the student will learn more information. In other words, distillation is the process in which the teacher teaches the knowledge he has learned from the sample data to the students.

When studying, students have two strategies:

1. Only use the output results of the teacher, that is, only learn the teacher.
2. Using the teacher's output results and the original labels of the samples at the same time. It can be understood as critical learning of the teacher. Generally, teacher's output is given greater weight.

Taking Logit Distillation proposed by Hinton as an example, the learning process is as shown in Fig. 4.14:

It makes improvements to the softmax function:

$$q_i = \frac{\exp(z_i / T)}{\sum_j \exp(z_j / T)}$$

There is one more parameter T (temperature) than the previous softmax. The larger T is, the smoother the probability distribution will be, and the more generalization information of the teacher model can be learned. T can try between 1 and 20.

The training data is input into the teacher model and the student model, respectively, and predictions are generated respectively. The teacher's output soft labels and the student's output student predictions are compared, and a part of the loss is calculated. On the other hand, the hard prediction output by the student is compared with the original label hard label of the sample, and another part of the loss is also calculated. The two parts of the loss are weighted and added to produce a complete loss, which is then back propagated and the weight of the student model is updated.

The core of Hinton's Logit Distillation is the design of the loss function. This idea can be applied to more models, such as Distilled BiLSTM, DistilBERT, BERT-PKD, TinyBert, MobileBert, MiniLm, etc.

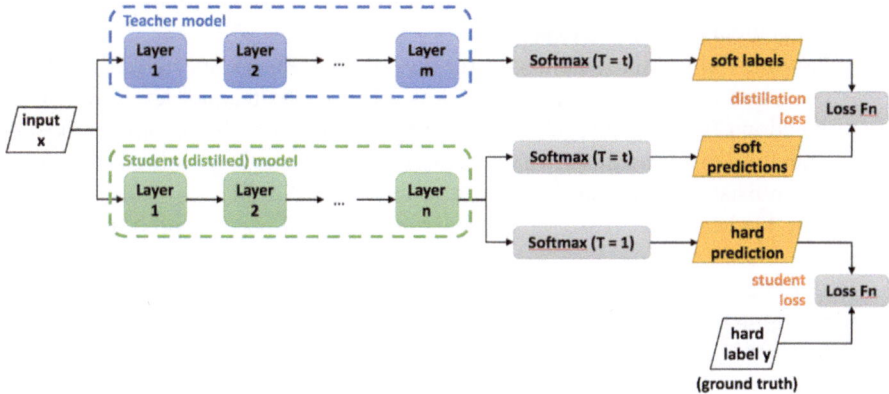

Fig. 4.14 Model distillation process (Intel labs, 2023)

Model distillation, as a means of model compression, is a relatively effective method to reduce model size. Its advantage is that it is very flexible and can be easily migrated from one model to another; however, it also has its own shortcomings, that is, it also requires additional training, which puts certain requirements on data and time. In addition, even the student model needs to be carefully designed to achieve better distillation.

Finally, regardless of the method, neural network compression is a double-edged sword, and the higher the compression ratio, the worse the accuracy of the model. Using different pruning rates results in different degrees of sparsity. The higher the degree of sparsity, the model size will decrease proportionally, but the accuracy will also slowly decrease. When the sparsity reaches a certain threshold, the accuracy will fall off a cliff, making the model unusable.

Therefore, care should be taken to find a suitable balance in the process of using neural network compression.

4.2   Compression–Compilation co-Design

The principle of compression–compilation co-design (Liu et al., 2020) is to compress the deep neural network model and compile the compressed model executable at the same time. This collaborative approach can effectively optimize the size and speed of deep neural network models and can also greatly reduce the adjustment time of the compression process so that deep neural network models can be deployed on mainstream processors (e.g., CPUs/GPUs) of embedded devices and achieve real-time performance on most AI applications that were originally thought to only use special AI accelerators (e.g., ASICs/FPGAs) in order to achieve the effect of real-time operation. This approach allows embedded neural networks to run on mainstream processors，without the need for special AI accelerators. Its advantages are as follows:

- **Cost**: Developing dedicated AI accelerators is costly and adding them to existing systems incurs additional expenses.
- **Technology maturity**: Unlike general-purpose processors, specialized hardware is produced in smaller quantities, and as a result, the technology available for production often lags that of general-purpose processors. For example, most AI accelerators today are based on 28–65-nm CMOS technology, which has a much lower transistor density than the latest mobile CPUs or GPUs.
- **Speed**: Dedicated processors run slower than general-purpose processors due to older technology.
- **Ecosystem**: General-purpose processors have a well-established ecosystem (debugging tools, optimization tools, security measures), which makes the development of high-quality applications much easier than using special processors.
- **Time-to-market**: The development of AI accelerators typically takes years. Creating associated compilers and system software for newly developed hard-

ware accelerators further lengthens the process. Applications using such hardware often require the use of special APIs and meet a number of special constraints, which can extend the time to market for AI products.

- **Availability**: For all the reasons mentioned above, the use of a special processor is usually limited to the company that created it and its few close customers. As a result, AI applications developed for special processors can only be adopted by a limited number of devices.

Therefore, the compression–compilation co-design scheme is a low-cost, software-based implementation of embedded artificial intelligence. The following describes the principle of compression–compilation co-design software algorithm scheme.

4.2.1   Concept of Compression–Compilation co-Design

Compression and compilation are two key steps to adapt deep neural network models on general-purpose hardware and perform them efficiently. Model compression is a common technique to reduce the size of deep neural network models and increase their speed, which has been described in the previous chapters, and pruning and quantization are mainly used in compression–compilation co-design. Compilation refers to the process of generating executable code from a given deep neural network model. Essentially, compilation is the process of mapping high-level operations in a deep neural network to low-level instructions supported by the underlying hardware. The compilation process plays a key role in optimizing the code for efficient execution.

The principle of compression–compilation co-design is to complete the design of both compression and compilation components at the same time, and this synergy is reflected in three levels.

- **Needs/Preferences Level**: At this level, synergy refers to the consideration of the preferences or needs of one component when designing another. One example is that mainstream processors often prefer code with certain computational patterns. If this preference can be considered in the model compression step, a more suitable model can be created so that the compilation step works efficiently.
- **Perspective/Insight Level**: At this level, synergy refers to the taking of the perspective or insight of one component into a problem while dealing with the other. One example is the principle of composability or modularity, which has always played a crucial role in keeping both programming systems and compilers efficient and scalable.
- **Method level**: At this level, synergy refers to the tight integration of approaches from two components. For example, compilers can look for new deep neural network pruning schemes by automatically generating code, which can achieve speedups of up to 180x.

Therefore, the compression–compilation collaborative system is an organic combination of compressor and compiler, which fully embodies the principle of collaborative design. They implement pattern-based DNN pruning and pattern-aware code generation, respectively, and generate efficient DNN execution code through synergy.

It consists of two components:

1. Compressor, a pattern-based compression component that performs convolutional kernel pattern pruning and connectivity pruning.
2. The compiler, which executes the compiled components generated by the code, performs multiple effective optimizations on the compressed model based on the pattern.

4.2.2 Compressor

The compressor mainly uses the method of weight pruning to compress the model.

Weight pruning is the mainstream model compression technique. However, existing pruning methods are either incompatible with modern parallel architectures, resulting in long inference delays (e.g., unstructured fine-grained pruning) or severe accuracy degradation (e.g., structured coarse-grained pruning).

The compressor implements a new weighted pruning technique by introducing fine-grained pruning within a coarse-grained structure. Because the fine-grained pruning mode enables higher accuracy, the code generated using the compiler can be reacquired and hardware efficiency is guaranteed. This approach can offer the best of both worlds, superior to previous pruning methods at the algorithm, compiler, and hardware levels.

1. Defects in Existing Compression Techniques

The two important categories of DNN model compression techniques are weighted pruning and weighting quantization. Weight pruning can reduce the redundancy of the number of weights, as shown in Fig. 4.15, the two main methods

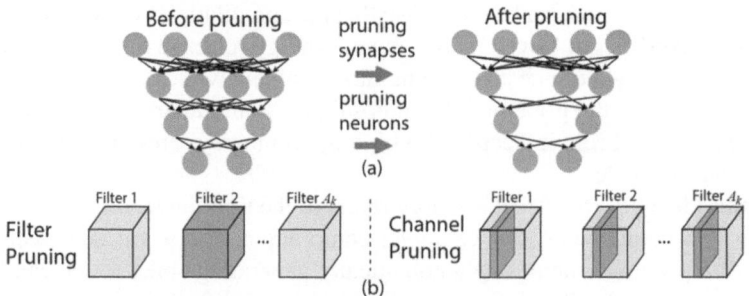

Fig. 4.15 (**a**) Unstructured weighted pruning (**b**) Two structured weighted pruning (Source: arXiv:2003.06700)

of weight pruning are: (1) unstructured pruning; (2) Structured pruning. These two methods generate irregular and regular DNN compression models, respectively.

Unstructured pruning: In this method, arbitrary weights can be pruned, so a high pruning rate can be achieved. However, due to the irregularities in computation and memory access, unstructured pruning can cause problems for compiler code optimization. Similarly, because pruned models are stored in an indexed sparse matrix format, unstructured pruning does not match the hardware accelerator and often results in GPU and CPU performance degradation.

Structured pruning: This method can generate a regularized weight matrix. The diagram above illustrates two representative structured pruning schemes: **filter pruning and channel pruning**. Filter and channel pruning can be considered equivalent, as the filter in pruning layer k is equivalent to the corresponding channel in pruning $k + 1$ layer. Filter/channel pruning is compatible with the Winograd algorithm. Due to structural rules, GPU/CPU running this pruning usually bring more significant speedup. However, structured pruning can result in a significant loss of accuracy.

Opportunity: Both pruning options have their advantages and disadvantages. In unstructured pruning, any weight can be pruned, which can be considered as a fine-grained approach, and in structured pruning, where the weights of the entire filter or channel are pruned, it can be considered as a coarse-grained approach. It would be nice if there was a way to combine high-accuracy unstructured pruning with hardware-friendly structured pruning.

To achieve this, we have introduced a new pattern-based pruning method *that* performs fine-grained pruning within a coarse-grained structure.

2. Pattern-Based Pruning

Pattern-based pruning techniques include convolution kernel mode pruning and connectivity pruning, which are both flexible and regular. Flexibility enables efficient model pruning, regularity enables compiler optimization, and hardware acceleration can be further leveraged.

Convolution kernel pattern pruning is shown in Fig. 4.16. For each convolution kernel in the filter, a fixed number of weights are pruned, and the remaining weights (the white cells in the figure) form a specific "pattern". The example in the figure retains four non-zero weights in the 3 × 3 convolution kernel. This method can be generalized to other sizes of convolution kernels as well as fully connected layers. Each convolution kernel can be flexibly selected from a set of predefined patterns.

At the theoretical and algorithmic level, the ideal convolution kernel pruning shape (non-square) matches certain connection patterns in the human visual system. As long as the appropriate mode is selected for each convolution kernel, the accuracy of the neural network can be improved. At the compiler level, the compiler rearranges and generates code, repeating convolution kernels with the same pattern

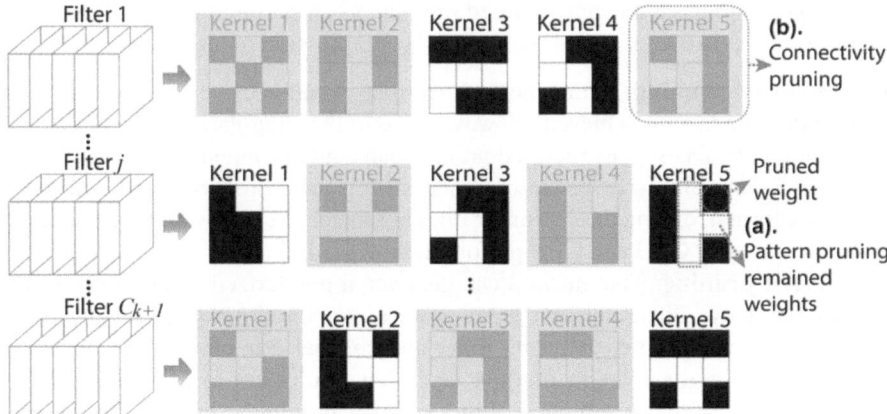

Fig. 4.16 Convolution kernel mode pruning (source: arXiv:2003.06700)

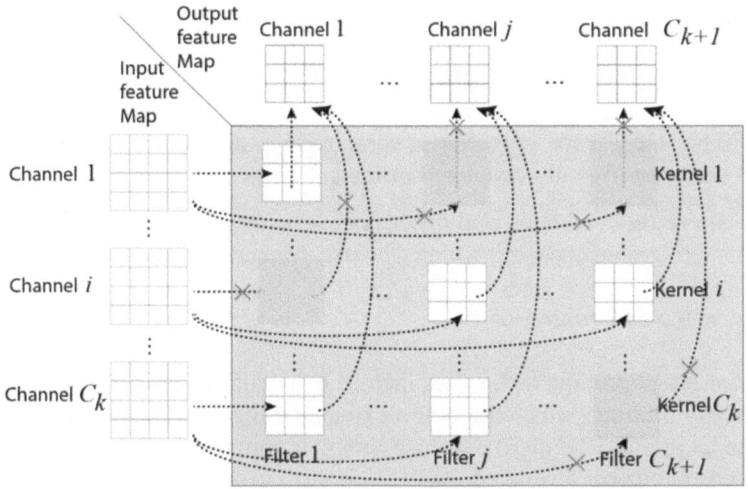

Fig. 4.17 Connectivity pruning (source: arXiv:2003.06700)

at the filter level and at the convolutional kernel level, maximizing instruction-level parallelism. At the hardware level, these modes are suitable for SIMD architectures in embedded processors and are suitable for both CPUs and GPUs.

Connectivity pruning is shown in Fig. 4.17. The key is to sever the connection between some of the input and output channels, which is equivalent to removing the corresponding convolution kernel. This method is combined with convolution kernel mode pruning to achieve a higher magnification of weight pruning.

At the theoretical and algorithmic level, connectivity pruning matches local hierarchical computations inspired by the human visual system. It is more flexible than filter and channel pruning and can achieve higher accuracy. At the compilation-level

Table 4.1 Comparison of progress and hardware acceleration rate of different pruning methods

Pruning method	Accuracy				Hardware acceleration			
	Highest	Slight loss	Moderate loss	Severe loss	Highest	High	Middle	Low
Unstructured	X							X
Filters/channels				X	X			
Convolutional kernel mode	X				X			
Connectivity		X				X		

hardware level, the deleted convolution kernels and related computations are grouped by the compiler using the sorting function without affecting other computations, thus maintaining parallelism.

Table 4.1 compares different pruning schemes, assuming that the pruning rate is the same, and qualitatively compares the accuracy and hardware acceleration of different pruning schemes.

The convolution kernel mode pruning has the highest accuracy and hardware acceleration rate, and the connectivity pruning also takes into account both the accuracy and the hardware acceleration rate.

Based on convolution kernel pattern pruning and connectivity pruning, the compressor first designs a set of patterns for each convolution kernel to select, and then performs filter pruning and connectivity pruning based on this pattern set. This results in a high pruning rate without sacrificing accuracy.

4.2.3   Compiler

The compiler performs multiple effective optimizations on the compressed model based on the pattern. First, the DNN model is converted into a computational graph, and this graph is optimized in several ways. Hierarchical design and optimization continue based on these optimizations, including high-level and fine-grained DNN hierarchical representation (LR), convolution kernel reordering, load redundancy depletion, and automatic parameter tuning. All these designs and optimizations are generic and work with mobile CPUs and GPUs. This component generates optimized execution code as well as a DNN model, with weights stored in a compact format.

Fine-grained hierarchical representation (LR) of DNN provides a high-level representation method that allows us to perform routine optimization of DNN models with a variety of resources. LR includes information about patterns and tuning. The optimization of the compiler relies on a series of improvements to LR to generate compact models and optimized execution code.

Convolution kernel and output channel reordering solve the two challenges caused by patterned pruning, namely dense control flow instructions and thread divergence and load imbalance, by combining convolution kernels of the same length and pattern. Due to the relatively limited number of convolution kernel patterns, convolution kernels with similar patterns can be grouped with appropriate convolution kernel reordering, which significantly reduces the control flow instructions and improves instruction-level parallelism. In addition, if different threads process different output channels, the problem of thread dispersion and load imbalance can be correctly solved because the convolution kernels in each output channel have a similar amount of computational effort, thus enhancing thread-level parallelism.

The Compressed Weight Storage format is specifically designed for convolution kernel mode and connectivity pruning. Combined with the convolution kernel and output channel rearrangement, this compact data structure has a higher compression ratio than the traditional CSR (Compressed Sparse Row) format.

Load redundancy elimination: By analytically dealing with two register-level load redundancy issues during code generation during convolutional kernel execution, the memory performance challenge of pattern pruning based on convolutional kernel is solved. This is even more important given that the movement of data between memory and cache has been optimized with advanced data tiling techniques.

Parameter auto-tuning specializes in testing different configurations of key performance parameters, including strategies for placing data on various GPU/CPU memories, different data tile sizes, and round-robin arrangements for each DNN layer on each processing unit.

In summary, the compression–compilation co-design approach allows the compiler to treat the pruned convolutional kernel as a special pattern, which can not only achieve high accuracy and high compression of the model, but also effectively convert the convolutional kernel pattern into optimized code on the hardware accelerator, combined with parallel computing and multi-level cache hierarchies, its potential will be unleashed more efficiently.

4.2.4 Advantages of Compression–Compilation co-Design

Combined with general-purpose hardware, the compression–compilation co-design approach is even more energy efficient than dedicated hardware accelerators. For example, the deployment of a compression–compilation co-design framework on the Samsung Galaxy S10 smartphone significantly outperforms some of the hardware accelerators implemented on common ASICs and FPGAs. Taking Google TPU as an example, the energy efficiency is increased to 4.76 times, and the inference delay is reduced to 1/18.75, as shown in Fig. 4.18.

Of course, one of the reasons why this combination is more energy efficient is because smartphones are inherently ultraenergy efficient. Therefore, the following comparison is made with other software acceleration frameworks (which only use

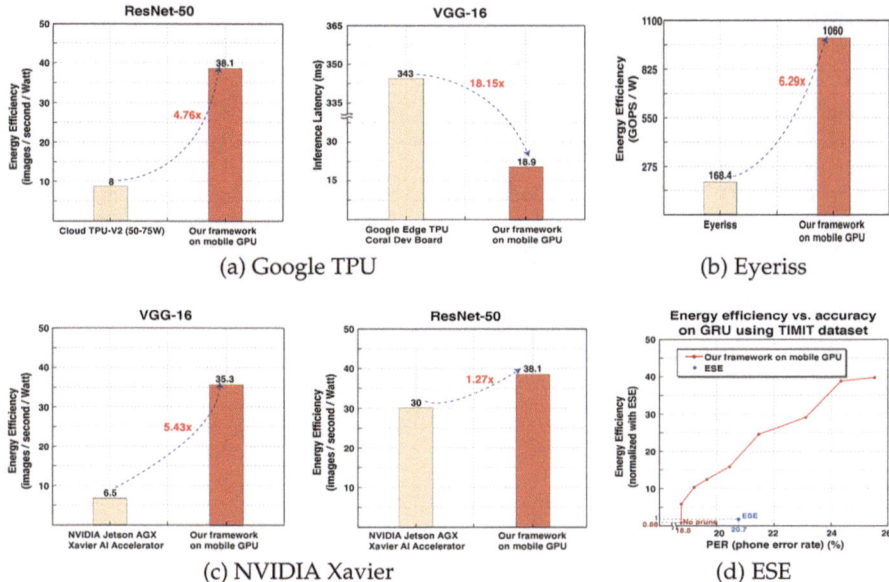

Fig. 4.18 Performance comparison between compression–compilation co-design and dedicated AI accelerator (source: arXiv:2003.06700)

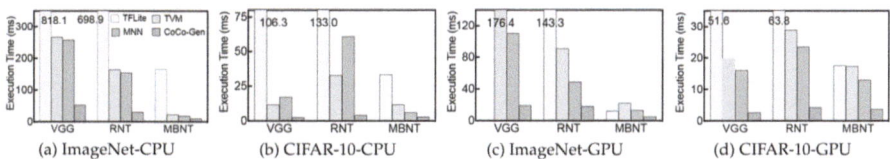

Fig. 4.19 Comparison between the compression and compilation co-framework and the model compression framework (source: arXiv:2003.06700)

model compression techniques) under the same hardware conditions, as shown in Fig. 4.19.

On the far right of each histogram in the figure is the inference execution time of the compressed compilation framework, which shows that the inference execution time of the compressed compilation co-design framework is significantly reduced compared to the framework that only uses model compression technology, whether it is for CPU or GPU.

In addition, it is worth noting that due to the high degree of flexibility of software-based solutions, the compressed compilation co-design framework can support all major types of neural networks and always maintain high performance, unlocking the full potential of embedded devices.

References

Han, S., Mao, H., & Dally, W. J. (2015). Deep compression: compressing deep neural networks with pruning, trained quantization and huffman coding. arXiv:1510.00149.

Hinton, G. E., Vinyals, O., & Dean, J. (2015). Distilling the knowledge in a neural network. arXiv:1503.02531.

Intel labs. (2023). *Knowledge distillation.* Retrieved from https://intellabs.github.io/distiller/knowledge_distillation.html

Liu, S., Ren, B., Shen, X., & Wang, Y. (2020). CoCoPIE: making mobile AI sweet as PIE—compression-compilation co-design goes a long way. arXiv:2003.06700.

Chapter 5
Framework for Embedded Neural Network Applications

Abstract This chapter introduces the fifth component to implement embedded AI, the embedded AI application framework. The energy efficiency of running neural networks in embedded devices can also be significantly improved through clever application-level optimizations. This chapter introduces the composition of this hierarchical cascade system, analyzes some key factors that can bring about efficiency improvements, and uses a case to demonstrate the cost efficiency improvements brought by this system. This chapter further extends this framework, distributes it to the cloud and devices, and proposes a third implementation model of embedded artificial intelligence: the device-cloud collaboration mode.

Keyword Device-cloud collaboration

Energy efficiency is the biggest bottleneck that neural networks cannot embed in IoT devices and mobile devices, which can be solved by developing more energy-efficient algorithms, or more energy-efficient hardware. But it can also be cleverly achieved through application-level optimization. This chapter introduces the hierarchical cascade system proposed by Bert Moons et al. (Moons et al., 2018). The system can improve system-level energy efficiency by leveraging the statistics of the input data.

In fact, there are already detection or classification systems that reduce the power consumption of always-on systems by adding a wake-up phase. They are commonly used in scenarios such as video surveillance systems, keyword recognition, or voice activation. In such a system, the first stage performs simple tasks at a low cost, rather than turning on the full but also more energy-intensive functions all the time. This significantly reduces the average running time of energy-intensive functions, which reduces the energy consumption of the entire system. For example, to implement the Siri voice assistant, the iPhone uses an always-on, low-power secondary processor (AOP) to trigger Siri, which consumes very little power. But once the user is asked to say "Hi, Siri," the AOP will wake up the main processor and analyze the user's voice with a more powerful deep neural network. The benefit of this approach is that it requires minimal processing to listen for and detect the "wake word," saving valuable battery power on the iPhone, and once it wakes up, it gives full play to

© Tsinghua University Press 2024

B. Li, *Embedded Artificial Intelligence*,

https://doi.org/10.1007/978-981-97-5038-2_5

the powerful processing power of the bionic neural engine. In addition, a tree-based multi-classification recognition system has been proposed, which has similar functions.

However, these systems can continue to be improved.

First, such systems usually have only two levels of hierarchy and no more, making them suboptimal in many cases. In addition, previous systems often did not explicitly utilize the statistical characteristics of the input data. More importantly, there was no framework for optimizing AI applications with a given accuracy in the direction of minimal energy consumption and applied to embedded processing scenarios.

By upgrading the two-stage cascade system to a general multi-stage cascade system, the performance is two orders of magnitude higher than the former, while the accuracy is still equal because the system can make fuller use of the statistic information of the input data.

Figure 5.1 introduces this system.

In this system, multiple stages are hierarchical cascaded, and as the stages increase (the level increases), the complexity and energy consumption of the sub-tasks increase gradually. The early stages only process simple data, while the later stages process more complex data, so that the early stages play the role of filtering out the data. For example, in speech recognition scenarios, wake-up words (e.g., "Alexa," "Hi, Siri," "Coco," etc.) are identified early in the hierarchy using inexpensive classifiers, thus avoiding the need for expensive stages to work all the way later. Therefore, while the later stages are more expensive, they are not used too much due to the discriminatory power of the earlier stages. This architecture combines the advantages of linear cascade and tree-based topology. As with the linear cascade method, most of the negative samples are eliminated before the final stage, saving power consumption for the entire system. Like the tree-based approach, it supports multi-classification problems. At the same time, if an error occurs in the early stages, it can still be corrected in the later stages. This system can be seen as a general-purpose framework for embedded AI applications that minimizes overall energy consumption and computational costs while maximizing the accuracy of the system.

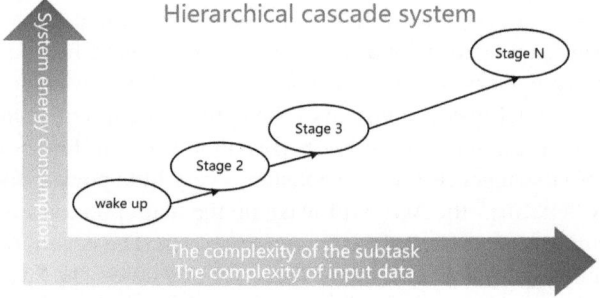

Fig. 5.1 General hierarchical cascade system

5.1   Composition of the Hierarchical Cascade System

A hierarchical cascade system is a general-purpose, multi-stage wake-up system. It minimizes the overall computational cost of the classification system without sacrificing performance. This is done by building a hierarchy of functions that are optimized for system-level energy efficiency: modules that are constantly increasing in function and cost are connected and jointly optimized. Hierarchies are more effective than single-stage systems, if the early stages are cheaper and the data can be adequately filtered into the more expensive later stages, and there are not too many mistakes that can't be recovered from in the later stages.

The whole system is divided into N stages, as shown in Fig. 5.2. It is assumed that the final task of the system is a complex multi-classification task. Usually, the first stage is used to identify the background, also known as the wake-up phase.

A typical hierarchy starts with a simple binary classification that removes the most obvious negative samples, such as background images or sound noise. If a positive sample is detected, i.e., a meaningful image is found, the next stage is activated for more precise classification. As with the first stage, in the later stages, only samples that were tested as "positive" in the previous stage are sent to the next stage. Throughout the framework, positive classes are referred to as "pass-on classes," and each next stage performs more complex classification tasks. Due to the higher complexity and performance, the cost of each stage increases significantly in a non-linear fashion.

To build a hierarchical cascade processing system, several alternatives were built and trained for each stage of the layering, each with a different balance of cost and accuracy. Each stage is independently trained on a dataset that simulates specific subtasks of the entire problem. Then, from a system perspective, the entire system is automatically optimized based on the performance of each stage, resulting in each accuracy or recall with minimal complexity or cost.

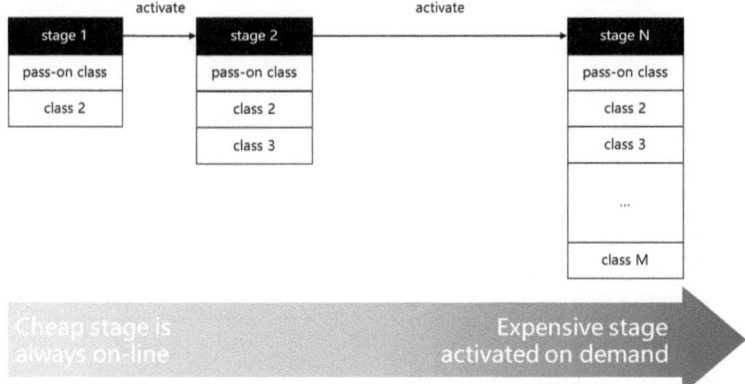

Fig. 5.2 Basic hierarchical classifier system with *N* stages and *M* categories

5.2   Efficiency of Hierarchical Cascade System

The previous section presents a theorized model of a hierarchical system. So, what is the effect of the statistical distribution of input data on the composition of this system? How many stages are optimal, and how much is the efficiency of the hierarchical cascade system compared to the original secondary wake-up system?

By studying a generic hierarchical cascade system that can classify 256 output categories, I it can be found that for uniformly distributed input data, it is optimal to include four stages in the model. For input data with a skewed distribution, such as most speech or image data, a higher number of stages are required in the lowest cost system.

Figure 5.3 shows the relationship between the number of stages and the overall cost for input data with different statistical distribution characteristics:

For uniformly distributed input data, the four-stage system achieves the ideal performance indicators. Six stages are required for moderately skewed input data and eight for highly skewed input data. For extremely skewed input data, the situation is more complicated, but at least six stages are needed for the system to reach the ideal performance indicators.

The reason for this is that in input data with a skewed distribution, the probability of certain words or certain types of images appearing is much greater than other words or images. Therefore, a simple classifier can be used to eliminate these easily recognizable classifications in the hierarchical structure at a lower cost, such as

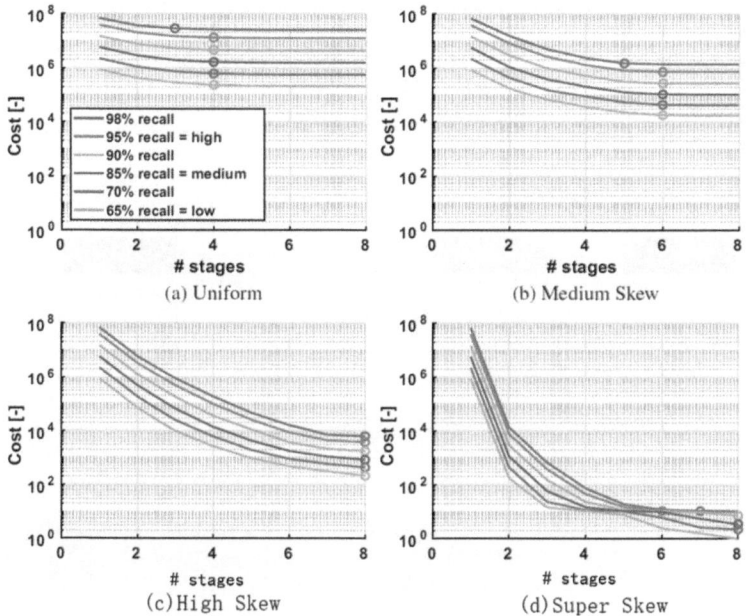

Fig. 5.3 Diagram of the number of phases versus the overall cost

"noise" or "background image. Words or images with a lower probability of occurrence are classified as "positive" at an earlier stage and are therefore passed on to the next stage for more precise recognition. In this way, the data passed to the last stage is minimal in probability. While it has the highest cost, it has the lowest probability of being activated and therefore accounts for the least of the total cost.

Therefore, the more skewed the statistical distribution of the input data, the better the cascade system with more stages. Of course, given the non-linear increase in the cost of a classifier at a higher stage compared to the cost at a lower stage, the increase in stages is not unlimited.

With the above mechanism, the efficiency of deep hierarchical cascade is three orders of magnitude higher than that of a two-level (wake-up) architecture with the same accuracy.

5.3   Hierarchical Face Recognition System

To illustrate the power of the developed method in a real-world system, we apply it to an actual hierarchical face recognition system using a convolution neural network (CNN). In the most basic method, such a system scans small windows at different scales of larger input images. Large-scale neural network-based face recognition is very expensive (1–2 mJ of energy per 250×250 window), especially in high-resolution images that need to process many windows, where the energy consumed increases proportionally to the square. Because input data is often statistically biased, it makes sense to build a processing hierarchy to take advantage of this property and reduce the average sample cost for the entire system.

To illustrate this, consider the following scenario: a pyramid ratio/sliding window approach on a 30 fps Full HD video (1920×1080 pixels) with a window size of 256, a step size of 4, and a scale factor of 2. To achieve real-time performance at 30 fps, more than three million windows per second (i.e., 100,000 windows per frame) need to be evaluated, but most windows should be evaluated as background images. Assuming that the evaluation consumes 1 mJ per window, a total of 3 kW of power is required, which is clearly not feasible on embedded, battery-limited devices. Since distinguishing a face from a background is a very simple task that can be performed in a 32×32 window and consumes only about 1 μJ of energy, such detectors can be used as a wake-up stage for more sophisticated and costly face recognizers.

Only when the face detector detects a face, the subsequent more expensive face recognizer is used. If not, the system will move to the next window, greatly reducing the average cost per window. While it is clear that a hierarchical system can bring huge benefits to this face recognition system, it is still unclear how many stages should be used in the hierarchy. If there is still a statistical bias between different face categories (e.g., the owner of the device appears much more frequently than other faces), you can add an intermediate stage to take advantage of this bias to further reduce the activation rate of the expensive final stage. Most importantly, the energy settings at all stages should be optimally adjusted to achieve minimal system energy consumption.

To investigate this, Fig. 5.4 outlines the wake-up method in this N-level hierarchical system that distinguishes between 100 faces. Only when the window is classified as a "pass-on-class" does the window pass to the next stage in the hierarchy. In all other cases, i.e., when an image is classified as "background," "owner," or another specific face, the window is considered to have been classified. Because the tasks in the previous phase are less expensive than the final tasks, the cost can be significantly reduced. In this case study, the entire hierarchy was optimized to minimize the average cost per sample while maintaining the overall face recognition accuracy. This analysis demonstrates that extending a typical two-stage wake-up system to a multi-level hierarchical cascade significantly reduces the cost.

In the example, the 100 face recognition systems are divided into four stages, which are used for face detection, owner detection, 10 face detection, and 100 face detection, where the last stage is a complete 100 face recognition machine. For each stage, 15 CNN models were trained with different complexity of downsampled dimensions of the input image, number of layers per network (network depth), and number of filters per layer (network width). Larger networks running on larger input images typically achieve higher recall and accuracy on the input dataset, albeit at a higher cost. The architecture and parameters of these networks are given in Table 5.1.

In these networks, the input size, the total number of layers (total depth), and the width of these layers are varied. The smallest network takes 32 × 32 RGB downsample as input and classifies it using four layers of CNN with four filters per layer. The largest network takes 128 × 128 RGB as input and classifies it using a 7-layer network with up to 1024 filters per layer. All networks are trained on a subset of 100 faces in the VGG FACE-2 dataset and used batch normalization and random data augmentation to prevent overfitting. LeakyRelu is used for all middle-layer activation functions, and softmax is used for Dense (fully connected) layer activation functions.

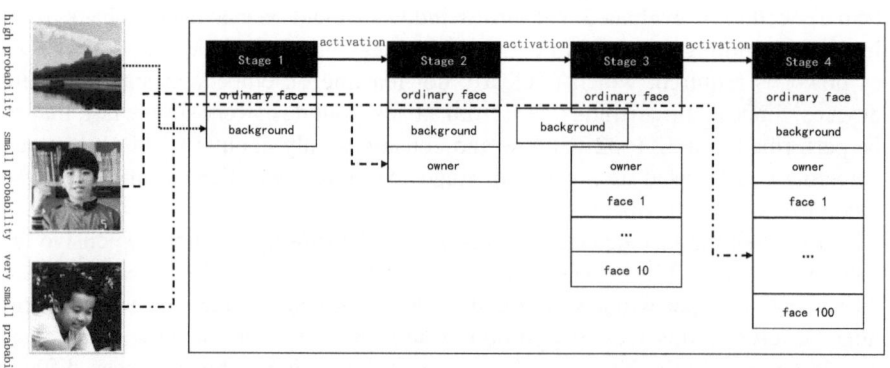

Fig. 5.4 Example of a hierarchical face recognition system. Backgrounds and owners are more likely to appear than other faces, detected using cheaper classifiers

Table 5.1 Hierarchical face recognition neural network topology and parameters

Block	Dimension	Kernel	Stride	# Layers	Width
Input layer	32-64-128	3-5-7	1-2-4	1	4-256
Block A	32	3	1	1	4-256
MaxPool	32	2	1	1	–
Block B	16	3	1	1-2	4-512
MaxPool	16	2	1	1	–
Block C	8	3	1	1-3	4-1024
MaxPool	8	2	1	1	–
Dense	$16 \times \text{width}_C$	–	–	1	–

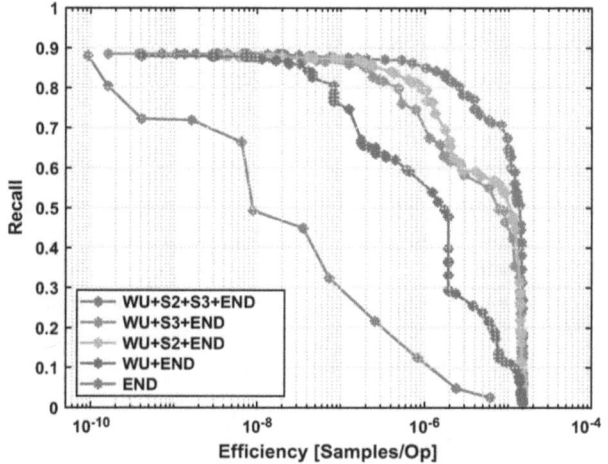

Fig. 5.5 Efficiency-recall curves of different hierarchical cascade structures

To evaluate the performance of the hierarchical cascade structure, comparing the efficiency and recall of several different hierarchies (from one stage to a maximum of four stages), the efficiency of the cascade system is defined as the number of samples processed per operation, i.e., the number of samples/operations, and the structure is optimal if the same efficiency achieves the highest recall. Figure 5.5 illustrates the efficiency-recall curve for these hierarchies when implementing 100 face recognition, with the upper right corner being the optimal. Obviously, the structure using only stage 4 is the least efficient. Similarly, the two-layer structure with only the wake-up stage and stage 4 is not optimal, and it requires 1 to 2 orders of magnitude more operations for the same recall than the optimal four-layer structure. However, it can be shown that the performance gains in deeper hierarchies beyond four stages will be less significant, especially when the maximum latency of the system is considered.

5.4   Device-Cloud Collaboration Mode

The hierarchical cascade system introduced above can reduce the energy consumption of neural network detection systems at the application level rather than at the algorithm or hardware level. This hierarchical cascade system can be implemented across devices. If the embedded AI chip can run all stages, especially the classifier in the final stage, this architecture can be implemented locally in the embedded device. But if this is not the case, then you can consider a distributed implementation approach, where the embedded device implements the first few stages of classification, and the cloud computing center is responsible for the higher stages of classification. The simplest case is the first stage of embedded AI: the cloud computing mode, where Amazon speakers are only implemented locally "Alexa" wake-up word recognition, and all other advanced speech and semantic recognition are implemented in the cloud. The complex situation can be a face recognition camera for access control. In most cases, it uses very low-power consumption to identify whether someone is approaching. When it finds someone approaching, it first uses a low-stage classifier. Identify whether it is the owner or a frequent visitor, otherwise a high-level classifier will be called to determine which stranger it is. Considering that no camera can accommodate 1 billion facial features, this last step should be left to the cloud.

This is the third mode of embedded AI, the device-cloud collaboration mode, as shown in Fig. 5.6.

In this mode, embedded AI chips play two roles:

1. Divide input data into two categories: common inputs and rare inputs.
2. Perform AI inference on common inputs.

The cloud computing center is responsible for handling rare inputs.

AI applications are cross-device and cloud-based. It is responsible for integrating the AI inference results locally on the device and the cloud and responding to inputs.

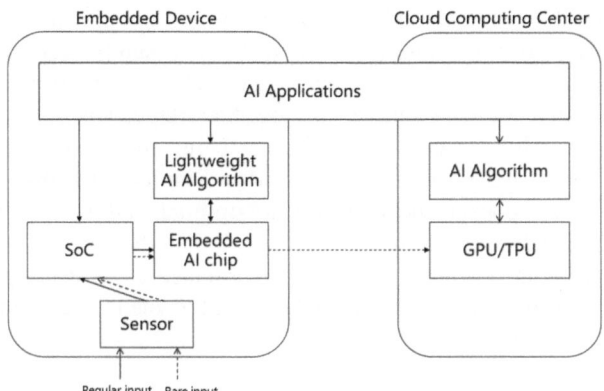

Fig. 5.6 Device-cloud collaboration mode—the third mode of embedded AI

It is called collaboration mode because in this mode, for most inputs, the cloud is not necessary, and embedded devices can work independently. The help of the cloud is only needed occasionally. In fact, it can also be used in this way, where inputs that require real-time response are handled by embedded devices, and inputs that do not require real-time performance are sent to the cloud. This combines the advantages of the local mode and the cloud computing mode and overcomes their respective shortcomings. It is a method that has the best of both worlds.

Reference

Moons, B., Bankman, D., & Verhelst, M. (2018). *Embedded deep learning: Algorithms, architectures and circuits for always-on neural network processing 1st ed. 2019 edition.* Springer.

Chapter 6
Lifelong Deep Learning

Abstract This chapter introduces an important future development direction of embedded AI, lifelong deep learning. By analyzing the shortcomings and causes of traditional deep learning, we clarify the goals and characteristics of lifelong deep learning and explore some methods to implement lifelong deep neural networks, such as dual learning systems, real-time updates, memory merging, and adaptation to real scenarios. Finally, the advantages brought by the combination of lifelong deep neural network and embedded AI are summarized, such as autonomous learning, federated learning, etc.

Keywords Lifelong deep learning · Lifelong neural network

Deep neural network technology has advanced by leaps and bounds in recent years and has been applied to countless AI systems, including autonomous vehicles, industrial applications, search engines, computer games, health record automation, and big data analysis. But at the same time, current deep neural networks **are still not intelligent enough in the biological sense**. They are unable to answer questions that they have not prepared in advance and are completely unable to recognize and respond to new situations or environments that they have not been specifically programmed or trained in. This shows that deep neural networks still have serious limitations in terms of system capabilities, limiting their application in some hard real-time tasks, such as supply chain, logistics, and visual recognition in the military field, where complete details are usually unknown, and it is critical to quickly adapt and react to dynamic environments. And that's exactly the kind of task we want embedded devices to accomplish.

This is contrary to the way the human brain works, which learns every day or even all the moment, very quickly, without forgetting past experiences and starting over.

It is in this context that lifetime deep learning technology was born, which can quickly learn after model deployment without the need for large amounts of training, heavy computing resources, and massive data storage. It is a new direction of deep learning (Neurala Inc., 2019).

© Tsinghua University Press 2024

B. Li, *Embedded Artificial Intelligence*,

https://doi.org/10.1007/978-981-97-5038-2_6

6.1 Drawbacks of Traditional Deep Learning

Rooted in the theoretical work in the 1960s, today's deep learning and neural network algorithms have achieved great success in various industries and scenarios. This large-scale system simulates brain activity by simulating the interconnection of neurons and is trained to perform a series of specific tasks. These tasks include visual and auditory perception, motion control, and more abstract functions such as catching cyberattacks on servers, classifying financial data as fraudulent or legitimate, or identifying a piece of equipment as genuine, defective, or rusty.

However, when a deep neural network system makes an error when it encounters an untrained situation, it must be taken offline and retrained. This approach is expensive and time-consuming, not to mention the fact that starting to retrain while performing a real-time task will cause the task to fail. Deep learning systems are also plagued by a major problem known as "catastrophic forgetting." They "forget" what they have learned before they train with new data, and unless they merge old and new data and train for every eventuality, these systems running in a real-world environment are bound to fail at some point. This means that current deep neural networks are limited to specific cases with a narrow set of predefined rules.

This is due to the following circumstances. The training process and inference process of traditional deep neural networks are completely separated, and the model freezes after learning and does not have the ability to continue learning.

The limitation and cost of implementing deep neural network technology lies in its large-scale learning and training requirements, such as:

- Millions of training data samples.
- Thousands of representations for each example.
- Hundreds of models that need to be trained for optimal performance.
- Once a use case or data changes, training will be started all over again.

These requirements translate into significant costs on many fronts: data collection and annotation, data storage, running memory, and computing power. In addition, developers need to build a model to start training and come back a few days later to evaluate performance and adjust parameters. This means months of annoying and inefficient development cycles, often requiring multiple iterations to get the right results.

The resulting deep neural network is "dead" and cannot be inferred, which means that even a change in a single data point requires tedious retraining of the entire data set. Learning is not incremental, which means that new knowledge cannot simply be added and must be retrained from scratch.

The underlying reason behind this is that most of systems use the backpropagation algorithm invented in the 1980s. This algorithm gives neural networks the ability to learn from data, which is a far improvement over the previous way of pre-programming a function.

In backpropagation algorithms, data can be images, sounds, or even more abstract objects, but the principle is the same: the algorithm optimizes the network output by iteratively adjusting the weights of each neuron, completing the learning process of these images or sounds. This enables backpropagation-based AI systems to match or even surpass human performance in a growing variety of tasks, from chess games to camera intrusion detection.

However, this super performance is costly: backpropagation networks are very sensitive to new information and are susceptible to catastrophic interference. All weights in a neural network are important for it to behave correctly, and attempting to add new information to the network requires modifying these weights, which often greatly disrupts previous knowledge, leading to a problem known as "catastrophic forgetting." When it learns something new, old information is erased.

To mitigate the effects of this problem, the researchers took two steps:

1. Slow down the learning and change only a small number of synapses at a time but recomputing all the inputs and repeat the process in multiple iterations. There is a small but consistent reduction in training error after each iteration. Thousands or millions of iterations are often required.
2. After reaching the target performance, freeze the learning process to avoid the impact on the old information learned when new information is added.

This still does not completely solve the problem: the deep neural network can only be used to the level it was trained on, and it doesn't learn anything new as it is used day in and day out.

Finally, in addition to the long training time and the need for a high-performance server, backpropagation has an additional drawback: it requires to store all input data for the next retraining. For example, a deep neural network has been trained on 1000 images, but if it needs to learn 1 new image, it still needs to process all 1001 images and go through tens or millions of iterations. What happens if the first 1000 images are not or cannot be legally retained? The answer is that the neural network cannot be trained nor updated.

In summary, traditional deep neural networks require long training times, expensive computing resources, expensive bandwidth, and connections to transfer data for training, and expensive storage to save data for the next step of training. What's more, it can't learn in real time.

To completely solve these problems, deep neural networks with lifelong learning capabilities are about to emerge, and some people have proposed the concept of lifelong learning machines (L2M). In other words, it is lifelong machine learning. If it is realized by deep neural networks, it is called Lifelong Deep Neural Network (Lifelong DNN), and its related algorithm is called Lifelong Deep Learning.

6.2 Goals of Lifelong Deep Learning

The goal of lifelong deep learning is to develop new machine learning methods that enable systems to continuously learn as they operate and apply previous knowledge to new contexts. Current AI systems only use pre-programmed or trained content for calculations, and they cannot learn from data inputs during execution or adapt online to the changes they encounter in a real-world environment.

It is hoped that new machine learning methods will be developed that will allow systems to learn from previous experiences and apply previous knowledge to new situations, just like the human brain.

Current machine learning systems are "not intelligent in the biological sense." The system does not have the ability to adapt to situations where it is not trained or programmed. This limits the effectiveness of AI systems in projects such as supply chain, logistics, and visual recognition, where the ability to react quickly and adapt to dynamic situations is paramount, as full details are often not known beforehand.

Instead of spending money to reprogram or retrain these systems, people want to develop a system that can learn like animals and humans learn through experience.

Therefore, the goal of lifelong deep learning is to develop fundamentally new machine learning mechanisms that continuously improve performance through experience and quickly adapt to new conditions and dynamic environments, so that the system can continuously learn as it performs and applies previously learned information to new situations like a biological system, that is, the environment is the training set.

Such a system is undoubtedly safer and more powerful, can be applied to some hard real-time tasks, can quickly adapt to unforeseen situations and changing tasks, and continuously improve performance through experience throughout the life cycle of the system, as shown in Fig. 6.1.

In the figure, the light-colored curve is a traditional AI, which does not increase its performance after training, and when it encounters unexpected circumstances, its performance drops sharply and cannot be recovered. Lifelong deep learning, on the other hand, can continue to learn after being deployed in the field, and continue to improve its performance, when encountering unexpected situations, it can quickly adapt to the changing environment, and its performance can rise after a short decline, even exceeding the pre-training level. This behavior is the same as that of a real biological brain (the human brain).

It is very difficult to implement such a system. It's easy to write agent behaviors that can perform specific tasks, but doing so prevents the algorithm from learning

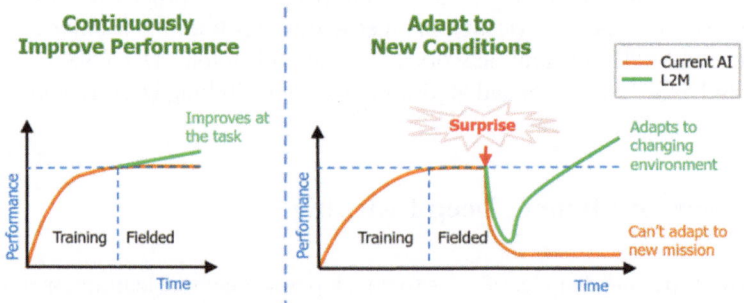

Fig. 6.1 The goal of lifelong deep learning

the task, precluding the possibility of adapting the behavior to other tasks or situations. This is at the heart of the problem that creating a lifelong deep learning algorithm is designed to solve. People can't help but want to use existing technology to implement specific behaviors, because we all know that seemingly complex behaviors can be hard-coded and often produce excellent results. However, this is completely contrary to the goal of lifelong deep learning, where the system figures out how to complete a task and, based on previous learning, subsequently completes another task more easily. And the behavior itself is a secondary factor in the development of this system, especially in the beginning.

6.3   Characteristics of Lifelong Deep Learning

Lifelong deep learning should have the following characteristics.

1. Balanced network: The possible implementation of a lifelong deep learning system is a plastic nodal network (PNN) rather than a fixed, homogeneous neural network. While malleable, PNN networks must be balanced with the hard rules that govern their operations. If the rules govern the PNN too strongly, the PNN is difficult to learn, but without structure, the PNN cannot operate at all.
2. Selective plasticity: Lifelong deep learning systems must also maintain a balance between plasticity and knowledge access. Fixed neural networks can't learn, and neural networks that vary too much can't retain knowledge. Hyper Turing computation has the inherent property of minimizing resource usage, i.e., it uses only the resources needed in any computation step. This property may limit changes to areas of nodes that are not directly involved in a particular computation, thus gaining a balance between knowledge acquisition and storage. PNN can only exist in the context of the entire system, which contains algorithms and is secure. The same trait can also be seen in humans: no matter how hard we try, we close our eyelids if an object is too close to our eyes. However, just like living organisms, the whole system must be in balance, and while hardwired instincts are beneficial, if we drive a car with that instinct, it would be disastrous if our hardwired instincts give different driving instructions than they are.
3. Goal-driven perception: Unlike fixed-data online learning, continuous learning involves fluctuating data distribution, uncertain concepts, and noise. During the computation, the label is most likely unavailable. But what's even more challenging is that there is no data to be trained in, and the system must actively select its inputs to adapt. Goal-driven perception introduces the possibility of dealing with the same situation in different ways depending on the goals of the system and the environment, as is the case with animals that are looking for food or mates, who will pay less attention to other opportunities. This form of perception results in a top-down data flow that combines bottom-up sensory data with an internal circulatory process. Goal-driven perception helps prevent overem-

phasizing less relevant inputs and instead focuses attention on critical inputs that require a response or adaptation from the system.

4. Adaptation to new situations: The driving force of the system is to adapt to new situations or introduce new goals. The system must know when to learn or present a new behavior. This requires that a lifelong deep learning system, unlike current machine learning, should not suffer catastrophic forgetting when learning new behaviors.

6.4 Inspiration from Neural Biology

If humans learn like traditional deep neural networks, we will only know what we are taught in school and never add new knowledge or insights. Lifelong deep neural networks should learn day in and day out, just like biological brains. Most of us have received some sort of standardized education, and in school we learn a range of important skills. As humans, even as our brains get older and become less agile, we still learn and improve our skills and knowledge every day. What's more, we can do it quickly: based on what we already know, it takes very little time to learn something new. Thankfully, if our brains can't do that, we can't survive as a species.

Lifelong deep neural networks take inspiration from the adaptive mechanisms of living organisms. Throughout their lifespan, biological systems exhibit an amazing ability to learn and regulate their structure and function while maintaining the stability of their core functions. This highly robust adaptive mechanism, developed over billions of years of evolution, is something that lifelong deep neural networks hope to emulate.

From neurobiology, lifelong deep neural networks are inspired by the following:

1. The lifelong deep neural network will change the architecture based on sensory input, combined with memories and system goals.
2. Memory and computation are part of each other, and they are different aspects of the same node connection.
3. Sensory input (e.g., visual, auditory) is first processed in a dedicated area of a lifelong deep neural network, and then associations are formed with other inputs and memory. The continuous adaptation mechanism is used to detect and identify stimuli in the natural environment, and nodes will be differentiated by algorithmic connection rules for processing different functions.
4. Biological neural networks have abundant signal systems, such as hormones, neuromodulators, calcium waves, neurotransmitters, etc., which can adaptively change neuronal characteristics at different scales (such as global, regional, local, etc.). Learning this property of the brain, lifelong deep neural networks can contain both local and differential signals (like neural action potential frequencies) and activate only those neighbors who can "listen" at that frequency. This architecture may require multiple connections when it might otherwise be

a single connection, greatly increasing complexity while limiting energy consumption.

5. Animals have some rules of operation from birth, i.e., instincts, some of which are present throughout life. Similarly, in lifelong deep neural networks, rules of behavior are combined with rules of learning to define how connections between nodes should be created, disconnected, weakened, and strengthened based on active memory, sensory input, and guidance from an external user, resulting in fixed functions.

6. Lifelong deep neural networks will not suffer from catastrophic forgetting based on the following properties, such as: differentiation (preventing nodes from changing globally based on arbitrary inputs); asynchronous updates (only some connections change); resource minimization (local updates instead of global changes).

The human brain is not the only object of lifelong deep neural network imitation, and can also take inspiration from the brains of lower animals, even those without brains, as long as they exhibit sufficiently good behavior. For example, cancer biology tells us the importance of cellular "nationality" within tissues, and the ability of axolotls to regenerate organs reveals a powerful adaptive mechanism. Natural systems of any size or type are worth learning from.

6.5   Implementation of Lifelong Deep Neural Network

Inspired by the neurophysiology of the brain, lifelong deep neural networks can mimic the ability of cortical and subcortical circuits to work together, dynamically adding new knowledge.

6.5.1   Dual Learning System

Since the early1990s, researchers have pointed out the existence of multiple complementary learning systems in the brain that can be combined to overcome the deficiencies of a single system.

For example, McClelland and colleagues proposed in 1995 that brain neocortical circuits have characteristics similar to those of backpropagation-based neural networks with highly distributed expression, high data compression, generalization, and slow learning. The hippocampus, on the other hand, has the characteristics of cyclic connection, sparse expression, and fast learning. They complement each other and form a **dual learning system**. In this way, backpropagation-based neural networks can overcome catastrophic forgetting.

The lifelong deep neural network constructs such a dual learning system on the basis of the traditional deep neural network, as shown in Fig. 6.2.

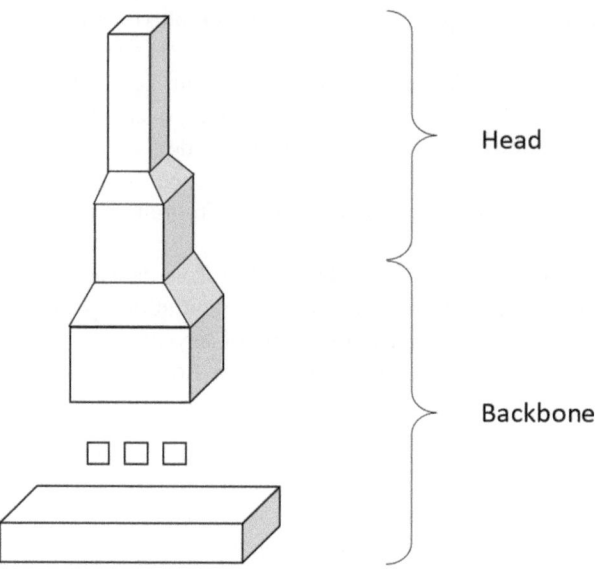

Fig. 6.2 Composition of a dual learning system

The dual learning system consists of two parts: the backbone and the head.

The backbone defines the quality of feature extraction, and its complexity often determines the speed and quality of inference. The same backbone can be used for different tasks, for example, ResNet is often used for classification, detection, and semantically segmentation of the backbone of neural networks.

The head defines the tasks performed by the neural network. The complexity of the head varies with the task. The head that performs classification is very simple, while the head that performs semantic segmentation is usually made up of many layers, upsampling the results to the appropriate resolution.

Fast real-time learning can be seen as a task, and while training a lifelong deep neural network affects the design and tuning of the backbone, its modifications are concentrated in the head. The head design includes many possible operations: the introduction of custom layers, new loss functions, special training process adjustments, etc. Lifelong deep neural networks are designed to use all these operations and build a fast learning brain based on the characteristics of the hippocampal system of the human brain (cyclic connection, sparse expression, and fast learning).

Lifelong deep neural network systems are based on the fact that backbone networks are excellent feature extractors. In order to connect the fast learning head to the backbone, depending on the situation, some of the topmost layers of the backbone network may be ignored, modified, or stripped away. For example, the original backbone network usually includes some fully connected layers, average pooling layers, which are used for the pre-training of the backbone neural network and enable the backbone neural network to make predictions directly but are not responsible for generating input for the new fast learning head of the lifelong deep neural network.

The quality of features and their dependencies on a specific set of objects are critical to the performance of a lifelong deep neural network. The better the feature set, and the more connections there are to generic "objects" (rather than specific objects), the higher performance a lifelong deep neural network will achieve for objects that are not pre-trained on the backbone. The quality of features depends on the architecture and training of the backbone, so it should be carefully investigated, compared, and even modified. In addition, adjusting the loss function and training process can also help to improve feature quality and object independence.

6.5.2   Real-Time Update

To be updated in real time, an important requirement for lifelong deep neural networks is that the dataset used for the pre-trained backbone network should have similarities in the feature space of the dataset used to train and test the lifelong deep neural network. For example, if the backbone network is pre-trained on the ImageNet dataset, the lifelong deep neural network performs well on images like it, but not well on images in the much smaller CIFAR-10 dataset. It performs worse on the MNIST dataset because the images are grayscale, and the feature spaces are very different.

However, if the backbone network has been pre-trained, the lifelong deep neural network learns new classes of objects quickly and accurately. For example, if the backbone network has been pre-trained on a subset of the MNIST dataset, i.e., training on 9 out of 10 digits (excluding 3), then the lifelong deep neural network will quickly learn the number 3 and will not forget the first 9 digits. The overall accuracy can reach 98.4%, which is 1.1% lower than the accuracy of an ordinary deep neural network trained directly on all ten numbers. However, the training time is greatly reduced compared to the original deep neural network. Using the same hardware and using the backpropagation algorithm, the training of the original network takes 6 h, while the training of a lifelong deep neural network can be completed in 0.4 s. That's 50,000 times faster learning.

In this way, the lifelong deep neural network achieves the goal of real-time updates.

6.5.3   Memory Merging

Lifelong deep neural networks solve a common problem in transfer learning through memory merging: overfitting.

In convention deep neural networks, fast training often leads to overfitting. This also exists in lifelong deep neural networks, but the impact is minimal, and the network recovers after several batches of training.

This effect is achieved because lifelong deep neural networks use a memory merging mechanism during training.

Memory merging takes advantage of the highly redundant nature of data representation in deep neural networks, where algorithms count weights in the head network, monitor weight changes, and detect outliers for each class. When the number of outliers in this category exceeds the threshold, the merge mechanism is automatically triggered.

Memory merging will shrink the cluster for each category by removing redundant weight vectors and setting them to initial values while preserving true outliers. The effectiveness of memory merging depends on feature quality: deleting weights in a sparse cluster will lose more information than deleting the same number of weights in an already compacted cluster. Memory merging also allows lifelong deep neural networks to learn more objects using the same amount of device memory.

One of the most interesting features of memory merging is that merging can be done across multiple lifelong deep neural networks. Of course, these lifelong deep neural networks need to have the same backbone network and a consistent annotation scheme (if different annotations are used for the same object, the system will be thrown into confusion). If these requirements are met, the heads of multiple lifelong deep neural networks can be merged into a single head, which will combine the knowledge of all the original heads.

This process can be likened to "brain fusion," as shown in Fig. 6.3.

It shares information across multiple devices, enabling federated learning. The fused "brain" can summarize more information and achieve better performance than each individual device. By updating the federated learning head to the device,

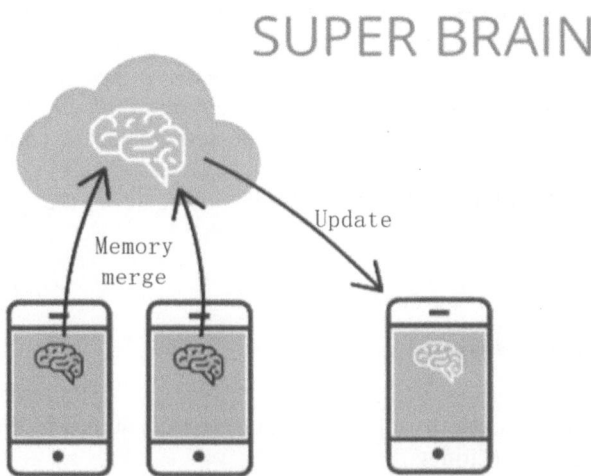

Fig. 6.3 Memory merging to achieve federated learning

the performance of the lifelong deep neural network on the device will also be improved.

6.5.4   Adaptation to Real Scenarios

The deployment of neural networks like lifelong deep neural networks, which allow for acquired learning, in real-world scenarios is very different from training in the lab on unpredictable input data. In a real-world scenario, new objects are likely to appear one after the other without being out of order, which is the rule, not the exception. In addition, it is highly likely that the network will need to perform inference on images that do not contain known objects at all. Traditional deep neural networks make what they think is their best guess when encountered, but this usually won't be the desired result. A better solution is to let the network learn this unknown object.

Lifelong deep neural networks can identify unknown objects through memory merging, and they can adapt to real-world scenarios by declaring new objects that are different from known class as "I don't know." This is achieved by using an implicit dynamic threshold to make predictions skewed, which requires that the recognition results be significantly concentrated on one category, rather than evenly distributed across multiple categories.

In other words, when the output of the neural network indicates that there is a clear winner in the known object category, it will recognize that the object belongs to this known category. However, when multiple different objects have approximate activation values and there is no clear winner, the system reports the object as unknown.

As an example, when the traditional neural network encounters the image of unknown class - "cow," it judges it as the closest class—"dog," but this self-righteous result is not as good as judging this image as "unknown."

This threshold is also very useful in training mode filter out well-learned object representations from cluttered representations that don't require attention and require no more learning. This provides the system with the opportunity to define and learn from objects that are classified as "unknown," increasing the intelligence of the system.

Figure 6.4 shows the learning process of the lifelong deep neural network in a real scene. The method is to input images of ten categories, 14 images for each category. During training, the images are input one by one sequentially, and after each image is input, verify.

The dark curve shows the Top-1 accuracy of the validation set after each image training. The vertical grid separates the training by category (14 training images per category). The light curve shows the validation results after forced memory merging after fully learning the images of each category. In most cases (8 out of 10), this merger further increases the accuracy of the system.

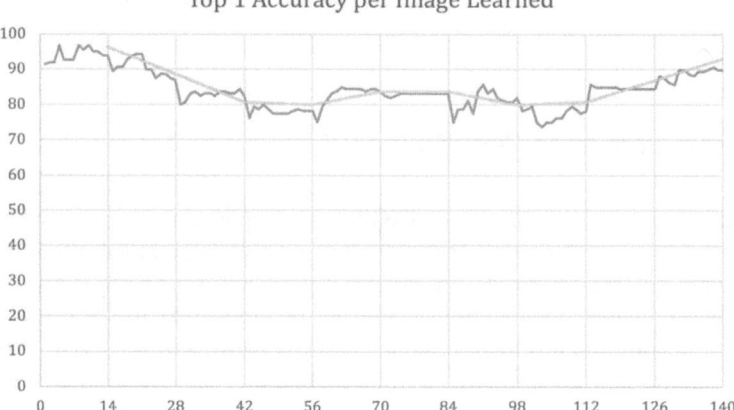

Fig. 6.4 Learning process of a lifelong deep neural network in a real-world scenario

Lifelong deep neural networks take deep learning to the next level by introducing techniques such as memory merging, real-time updates, and adaptation to real-world scenarios, solving the problem of catastrophic forgetting and allowing transfer learning and continuous learning.

6.6   Lifelong Deep Learning and Embedded Artificial Intelligence

Compared with traditional deep neural networks, the advantages of lifetime deep neural networks are enormous:

- The amount of data needed to build a robust network is greatly reduced.
- Training time can be drastically reduced: from days to seconds.
- Learning happens on the device, not on the server.
- No need to retrain when adding new knowledge.
- No need to store data.
- The system is more accurate because of continuous learning after deployment.

These advantages of lifelong deep neural networks make it possible to learn quickly and natively on embedded devices. These scenarios include smart terminals, security monitoring, supply chain monitoring, disaster and emergency response, drones, robots, and more. The technical characteristics of the lifelong deep neural network are updated in real time and adapt to real-world scenarios, so that embedded devices can learn from the environment after deployment, to better adapt to their unique environment. With federated learning, embedded devices can even achieve a kind of swarm intelligence, where each device can contribute what it has learned to form knowledge shared by the group. In this sense, lifelong deep neural network is a deep neural network suitable for the future development of embedded devices, and lifelong deep learning is an important direction for the future development of embedded artificial intelligence.

Reference

Neurala Inc. (2019). *How lifelong-DNNTM solves for inherent problems with traditional DNNs.* Retrieved from https://info.neurala.com/hubfs/Collateral/Neurala_LifelongDNNWhitepaper.pdf

Part II
Platforms

Chapter 7
Embedded AI Accelerator Chips

Abstract This chapter introduces the implementation of AI acceleration chips in detail. First, the advantages and disadvantages of several AI chip types such as GPU, DSP, FPGA, and ASIC are introduced. Then, several representative embedded AI chips are introduced, including NVIDIA Jetson, Intel Movidius, Google Edge TPU, XILINX DPU, and ARM Ethos NPU, Qualcomm Hexagon, etc. are taken as examples, and their working principles, internal structures, specifications, and application scenarios are introduced in detail. Finally, the above-mentioned main embedded AI accelerators are compared in terms of AI inference performance, power consumption, and inference performance per watt to facilitate embedded system developers to choose the appropriate AI acceleration chip according to their needs.

Keyword Embedded AI chip

7.1 Overview

Initially, AI hardware accelerators were designed to accelerate neural network calculations on servers. As these accelerators matured, their power consumption, size, and cost also gradually decreased. At the same time, artificial intelligence has become a key driver of edge computing, and the demand for AI inference on embedded devices, including mobile devices, is increasing. A series of AI accelerators suitable for embedded devices are beginning to emerge.

Embedded AI accelerators are mainly used to accelerate the inference of neural networks. These accelerators fall into several broad categories.

1. GPU: Graphics processing unit usually has thousands of convolution kernels and is suitable for large-scale parallel computing. It is currently the most widely used AI acceleration chip. Embedded GPU is a streamlined version of server GPU. Its advantages are mature technology, low cost, wide adaptability, and fast development speed. The disadvantage is that its performance is slightly inferior. Its representative is NVIDIA's Jetson series.

© Tsinghua University Press 2024

B. Li, *Embedded Artificial Intelligence*,

https://doi.org/10.1007/978-981-97-5038-2_7

2. DSP: Digital Signal Processor, which was originally used to perform matrix algorithms, so it can also be used for neural network calculations. However, the number of cores of DSP is insufficient, so its performance is not as good as that of GPU. However, in the embedded environment, DSP can play its strengths, such as fast speed and low-power consumption. The disadvantage of DSP is that it is specialized and generally developed for a specific task, for example, Qualcomm's Hexagon processor.

3. FPGA: Field Programmable Gate Array, which can implement neural network calculations at the hardware level and program circuits through software to quickly implement new functions and adapt to a certain extent. The rapid development of deep neural network technology. Its advantage is that the performance is higher than that of GPU, and the hardware can be programmed and can flexibly adapt to new neural network algorithms. Of course, its performance is lower than that of ASIC, and its flexibility still lags that of general-purpose processors. In addition, FPGA can only implement fixed-point operations, which requires the neural network model running in it to be quantized first, that is, reducing floating-point numbers to fixed-point numbers, which will lose a certain accuracy. Its representatives include XILINX's Zynq DPU.

4. ASIC: Application-specific integrated circuit, which can be customized and optimized for the special needs of neural network computing to maximize execution efficiency. Among the above-mentioned chips, ASIC has the lowest unit cost and the highest energy efficiency. Of course, the development cost of ASIC is high, and the cycle is long. With the rapid development of deep neural network technology, once ASIC is customized, it cannot be modified, and it is difficult to support new algorithms. The usual approach is to first implement a special neural network algorithm based on FPGA, and then use ASIC to reduce costs after the functions are solidified. Its representatives include Google's Edge TPU.

Table 7.1 summarizes the similarities, differences, advantages, and disadvantages of these methods.

The following introduces several representative embedded AI accelerators, including NVIDIA Jetson, Intel Movidius, Google Edge TPU, XILINX DPU, ARM Ethos NPU, Qualcomm Hexagon, etc., which are representatives of the above chip types.

Table 7.1 Embedded AI chip classification

Embedded AI chip type	Versatility	Performance	Power consumption	Cost	Development cycle
GPU	High	Middle	Higher	Middle	Short
DSP	Low	Middle	Low	Low	Middle
FGPA	Middle	High	Low	Low	Middle
ASIC	Low	Highest	Low	Lowest	Long

7.2   NVIDIA Jetson

When it comes to GPUs, NVIDIA is the undisputed market leader. Almost every public cloud provider offers NVIDIA-based GPUs, such as the T4, V100, and A 100 processors, as part of their infrastructure services. The company also sells DGX and EGX servers, which come with multiple high-end GPUs and are designed to run deep learning, high-performance computing, and scientific workloads.

NVIDIA has launched the Jetson series of embedded AI modules (NVIDIA Corporation, n.d.) based on server GPUs. High-performance, low-power NVIDIA Jetson systems deliver real-time artificial intelligence performance on embedded, small, energy efficient devices for processing complex data to rely on fast, accurate inference in network-constrained environments. In terms of programmability, they are 100% compatible with their enterprise data center counterparts. But these GPUs have fewer cores and consume less power than traditional GPUs used in desktops and servers.

NVIDIA Jetson also provides a supporting developer toolkit—JetPack SDK, which provides a unified software architecture for developing and deploying applications in various scenarios, thereby reducing software development costs. It has the necessary drivers, runtimes, and libraries to run machine learning and AI models at the edge. Data scientists and developers can easily convert TensorFlow and PyTorch models to TensorRT, a format that optimizes model accuracy and speed.

7.2.1   Introduction to Jetson Module

Jetson family includes a range of products that can be used in a wide range of sectors such as manufacturing, logistics, retail, services, agriculture, smart cities, and healthcare and life sciences to create breakthrough products. Among them, Jetson Xavier is a high-end module, Jetson TX2 is a mid-range module, and Jetson Nano is NVIDIA's most affordable GPU module ever. The Jetson Nano Dev Kit looks very similar to the Raspberry Pi, allowing enthusiasts, manufacturers, and professionals to build the next generation of AI and IoT solutions.

1. **Jetson AGX Xavier Series**

    The Jetson AGX Xavier series module is shown in Fig. 7.1.

    It enables higher levels of computing density, energy efficiency, and AI inference capabilities at the edge. Users can use the lower power and lower price Jetson AGX Xavier 8 GB module, configured for 10 or 20 W operating modes depending on the application, or use the Jetson AGX Xavier module and configure it for 10, 15, or 30 W operating modes. These modules are 10 times more energy efficient and 20 times more powerful than Jetson TX2.

2. **Jetson Xavier NX**

    Jetson Xavier NX is shown in Fig. 7.2.

Fig. 7.1 Jetson AGX
Xavier module

Fig. 7.2 Jetson Xavier NX module

It brings up to 21 TOP accelerated AI computing to the edge in a small module. It can run multiple modern neural networks in parallel and process data from multiple high-resolution sensors, a requirement for a complete AI system. Jetson Xavier NX supports all popular AI frameworks.

3. **Jetson TX2 Series**

   Jetson TX2 is shown in Fig. 7.3.

   It is built around the NVIDIA Pascal family of GPUs and features a variety of standard hardware interfaces to easily integrate it into a variety of products and form factors. The Jetson TX2 embedded module for edge AI applications comes in three versions: Jetson TX2, Jetson TX2i, and the less expensive Jetson TX2 4 GB.

4. **Jetson TX1**

   Jetson TX1 is shown in Fig. 7.4.

   It is the world's first module-based supercomputer, delivering the performance and energy efficiency required for visual computing applications. It is built on the NVIDIA Maxwell architecture with 256 CUDA cores, delivering over 1 TFLOP of performance to support embedded deep learning, computer vision, graphics, and GPU computing systems. The new design should use a Jetson TX2 4 GB to run neural networks at the same price, making the neural network twice as computationally efficient, or twice as energy efficient.

5. **Jetson Nano**

   NVIDIA Jetson Nano is shown in Fig. 7.5.

   It is a small, powerful computer suitable for embedded AI systems and the Internet of things, delivering the power of modern AI on a low-power platform. Get started quickly with the NVIDIA Jetpack SDK and a complete desktop Linux environment and start exploring a new world of embedded products.

Fig. 7.3 Jets on TX2 module

Fig. 7.4 Jetson TX1 module

Fig. 7.5 Jetson Nano module

7.2.2 *Jetson Module Internal Structure*

Taking Jetson Xavier NX as an example, the structure of the Jetson module (NVIDIA Corporation, 2019) is shown in Fig. 7.6.

Jetson Xavier NX includes an integrated 384-core NVIDIA Volta GPU (with 48 Tensor cores), a 6-core NVIDIA Carmel ARM v8.2 64-bit CPU, 8 GB 128-bit

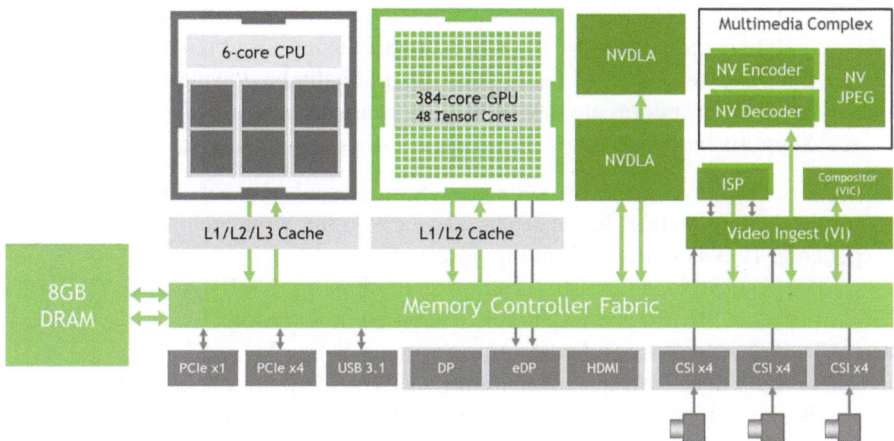

Fig. 7.6 Block diagram of the Jetson Xavier NX processor engine, including high-speed I/O and memory structures

LPDDR4x, 2 NVIDIA Deep Learning Accelerator (NVDLA) engines, a 4K Video encoder and decoder, a dedicated camera that can receive up to 6 high-resolution sensor streams simultaneously, PCIe Gen 3 expansion, dual DisplayPort/HDMI 4K displays, USB 3.1 and GPIO including SPI, I²C, I²S, CAN bus, and UART. The shared memory structure allows processors to freely share memory without generating additional memory copies (called ZeroCopy), which effectively improves the system's bandwidth utilization and throughput.

For Jetson Nano, the core GPU uses NVIDIA Maxwell architecture, while Jetson TX2 uses NVIDIA Pascal architecture.

Depending on the workload, the Dynamic Voltage and Frequency Scaling (DVFS) governor dynamically adjusts the operating frequency of the Jetson acceleration engine at runtime to reduce power consumption when idle.

The core of Jetson Xavier NX is the Volta GPU and NVDLA deep learning accelerator, which are explained below.

1. Volta architecture GPU

    The Jetson module uses the same Volta architecture GPU as NVIDIA's most powerful server-level AI processor. It is NVIDIA's first GPU using Tensor cores. This specially designed core provides higher deep learning performance than regular CUDA cores.

    Volta architecture are

    (a) CUDA computing capability version 7.0 supports concurrent execution of integer and floating-point operations.
    (b) High-speed bandwidth memory.
    (c) NVLink 2.0: A high-speed bandwidth bus between CPU and GPU and between multiple GPUs. Allows much higher transfer speeds than those achieved using PCI Express, up to 25 Gbit/s per lane.

(d) Tensor core: The tensor core is a unit that multiplies two 4×4 FP16 matrices and then adds a third FP16 or FP32 matrix using fused multiply-add operations and gets an FP32 result that can optionally be downgraded to an FP16 result. Tensor cores are designed to speed up the training of neural networks.

(e) Supports hardware video decoding of PureVideo feature set I.

With the above functions, Volta provides powerful AI processing performance for the Jetson module. In the Jetson Xavier NX, the Volta architecture GPU can generate 12.3 TOPS of computing power.

The Volta architecture is a common technology for NVIDIA GPUs (NVIDIA Corporation, 2018b) and will not be described in this book.

2. NVDLA Deep Learning Accelerator

NVDLA deep learning accelerator (NVIDIA Corporation, 2018a) provides a simple, flexible, and powerful inference acceleration solution. It supports a wide range of performance levels, from smaller, cost-sensitive IoT devices to larger performance-oriented IoT devices and can easily scale applications. NVDLA is delivered as a set of IP core models based on open industry standards:

(a) Verilog model, a synthesis and simulation model in RTL form.

(b) TLM SystemC simulation model can be used for software development, system integration, and testing.

NVDLA is a standardized open architecture. It is scalable and highly configurable, and its modular design maintains flexibility and simplifies integration. The NVDLA software ecosystem includes an on-device software stack, a complete training architecture for building new models incorporating deep learning, and parser software to convert existing models into an on-device usable form.

NVDLA Modular Architecture

NVDLA introduces a modular architecture designed to simplify configuration, integration, and portability; it exposes the building blocks used to accelerate the core's deep learning inference operations. NVDLA hardware consists of the following components:

- Convolution Core—Optimized high-performance convolution engine.
- Single Data Point Processor—A single point lookup engine for activation functions.
- Flat Data Processor—Flat averaging engine for pooling.
- Cross-Channel Data Processor—Multi-channel averaging engine for advanced normalization capabilities.
- Dedicated Memory and Data Reshape Engine—A memory-to-memory translation acceleration engine for tensor reshape and copy operations.

These components are independent and configurable independently. For example, systems that do not require pooling can remove the planar averaging engine entirely; or systems that require additional convolution performance can scale the performance of the convolution unit without modifying other units in the accelerator. The scheduling operation of each unit is handed over to a coprocessor or CPU, which is scheduled at extremely fine granularity, and each unit runs independently. This requirement for tight management scheduling can be implemented as part of the NVDLA subsystem ("headed" implementation) by adding a dedicated management coprocessor; or the functionality can be merged with a higher level driver on the main system processor ("Headless" implementation).

NVDLA hardware uses standard methods to interface with the rest of the system:

- A control channel that implements the register file and interrupt interface.
- A pair of standard AXI bus interfaces are used to interface with the memory.

The main memory interface is designed to connect to the system's broader memory system, including system DRAM. This memory interface should be shared with the system's CPU and I/O peripherals. The second memory interface is optional and allows connection to higher bandwidth memory that may be dedicated to NVDLA or a general computer vision subsystem. This heterogeneous memory interface option provides additional flexibility for scaling between different types of host systems.

The typical flow of inference begins with the NVDLA management processor (controller in a "headed" implementation or the main CPU in a "headless" implementation) sending down a hardware layer's configuration and an "activate" command. Multiple hardware layer configurations can be sent to different engines and activated simultaneously if the data are not related to each other (i.e., if there is another layer whose input does not depend on the previous layer's output). Because each engine has a double buffer for its configuration registers, it can also capture the configuration of a second hardware layer so that processing of the other hardware layer can begin as soon as the active layer completes. Once the hardware engine completes its active tasks, it issues an interrupt to the management processor to report completion, and the management processor begins the process again. This command-execute-interrupt process is repeated until inference for the entire network is complete.

NVDLA implementations generally fall into two categories:

- Headless: Unit—by-unit management of NVDLA hardware occurs on the main system processor.
- Headed: High interrupt frequency tasks are handed over to the supporting microcontroller (coprocessor) that is tightly coupled with the NVDLA subsystem.

The small system model in Fig. 7.7 shows an example of a headless NVDLA implementation, while the large system model shows a headless implementation. Small models represent NVDLA implementations for specialized devices that are more cost sensitive. Larger models feature the addition of dedicated control coprocessors and high-bandwidth SRAM to support the NVDLA subsystem. Large system models are more suitable for high-performance IoT devices that can run multiple tasks simultaneously.

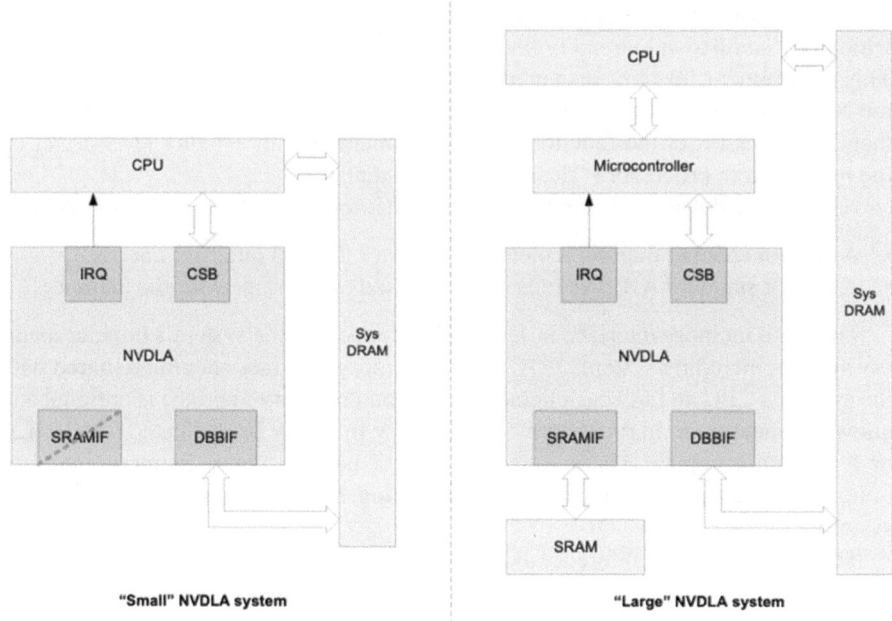

Fig. 7.7 Comparison of two possible NVDLA systems

Small NVDLA Model

Small NVDLA models bring deep learning technology into areas where it was previously unfeasible. This model is ideal for cost-sensitive IoT-type devices as well as artificial intelligence and automation systems that are very sensitive to cost, area, and power. With NVDLA configurable resources, you can save cost, area, and power consumption. Neural network models can be pre-compiled, and performance optimized, and large network models can be "cut down" and reduced in complexity. Conversely, a scaled-down NVDLA implementation allows the model to consume less storage space and reduce system software loading and processing time.

Such purpose-built systems typically perform only one task at a time, so it is usually acceptable to sacrifice some system performance while NVDLA is running. The context switching overhead associated with these systems is small and the main processor is not overburdened by handling large numbers of NVDLA interrupts. This eliminates the need for an additional microcontroller, and the main processor is sufficient to perform coarse-grained scheduling and memory allocation as well as fine-grained NVDLA management.

Typically, systems that follow the small NVDLA model will not include an optional second memory interface. When overall system performance is less important, the impact of not having high-speed memory channels is unlikely to be critical.

In such systems, system memory (usually DRAM) may consume less power than SRAM, so it is more energy efficient to use system memory as a computational cache.

Large NVDLA Model

High performance and versatility are primary considerations, and large NVDLA models are the better choice. Performance-oriented IoT systems can inference for many different network topologies, and therefore, these systems must maintain a high degree of flexibility. Additionally, these systems may perform multiple tasks simultaneously, rather than performing inference operations serially, so inference operations cannot consume excessive processing power on the host machine. To meet these needs, NVDLA hardware includes a second optional memory interface dedicated to high-bandwidth SRAM and the ability to interface with a dedicated control coprocessor (microcontroller) to reduce interrupt load on the main processor.

When included in the implementation, a high-bandwidth SRAM is connected to the fast memory bus interface on NVDLA, and this SRAM is used by NVDLA as a cache. Optionally, it can be shared with other high-performance computer vision components on the system to further reduce traffic accessing main system memory (Sys DRAM).

The requirements for the NVDLA coprocessor are generic, so there are a variety of suitable general-purpose processors to choose from (for example, the RISC-V based PicoRV32 processor, ARM Cortex-M or Cortex-R processors, or even internal microcontrollers). When using a dedicated coprocessor, the host processor still handles some tasks related to managing NVDLA. For example, while the coprocessor is responsible for scheduling and fine-grained programming on the NVDLA hardware, the host will still be responsible for coarse-grained scheduling on the NVDLA hardware, IOMMU mapping for NVDLA memory accesses (as needed), for input data and fixed weights on NVDLA memory allocation of arrays, and for synchronization between other system components and tasks running on NVDLA.

Hardware Architecture

NVDLA hardware architecture is shown in Fig. 7.8.

The NVDLA architecture can be programmed in two modes of operation: standalone mode and fused mode.

- Standalone mode. When run independently, each functional block is configured for when and what it executes, and each block independently performs the tasks assigned to it (similar to independent layers in deep learning frameworks). Standalone operations begin and end with designated function blocks performing memory-to-memory operations, reading, and writing from main system memory or dedicated SRAM memory.

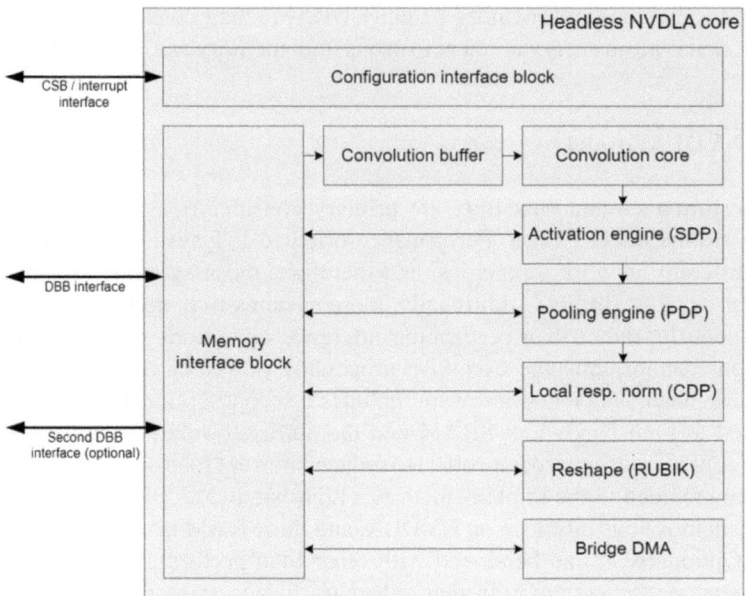

Fig. 7.8 Internal architecture of NVDLA core

• Fused mode. Fusion operations are similar to independent operations; however, some functional blocks can be assembled into a pipeline. Improve performance by bypassing round trips to memory, rather than having function blocks communicate with each other via small FIFOs (i.e., a convolution kernel can pass data to a single data point processor, and a single data point processor can pass data to flat data processing processor and then passed to the cross-channel data processor).

Interface

NVDLA has three main interfaces with the rest of the system:

• **Configuration Space Bus (CSB) interface**. This interface is a synchronous, low-bandwidth, low-power 32-bit control bus designed for CPU access to NVDLA configuration registers. NVDLA acts as a slave on the CSB interface. CSB implements a very simple interface protocol so it can be easily converted to AMBA, OCP or any other system bus with a simple shim layer.
• **Interrupt interface**. NVDLA hardware includes a 1-bit level driven interrupt. The interrupt line is set when the task is completed or an error occurs.
• **Data Backbone (DBB) interface**. The DBB interface connects NVDLA to the main system memory subsystem. It is a synchronous, high-speed, and highly configurable data bus. You can specify different address sizes, different data

sizes, and make different size requests based on the requirements of your system. The Data Backbone Interface is a simple interface protocol similar to AXI (and can be easily used in AXI-compliant systems).

The DBB interface has an optional second interface that can be used when a second memory channel is available. This connection is designed to be identical to the main DBB interface and can be used with on-chip SRAM that provides higher throughput and lower access latency. The second DBB interface is not required for NVDLA to operate, and systems that do not require this memory interface can save space by removing it.

Components

Each component in the NVDLA architecture supports specific operations that are integral to deep neural network inference. The following describes the brief functionality of each component, including the TensorFlow operations mapped to them. Although TensorFlow operations are used as an example, NVDLA hardware supports other deep learning frameworks.

Convolution

The convolution operation processes two sets of data: a set of offline-trained "weights" that remain constant at each inference and a set of input "feature" data that changes with the input to the network. The convolution engine exposes its parameters, allowing convolutions of many different sizes to be efficiently mapped to hardware. The NVDLA convolution engine optimizes the performance of the original convolution implementation. Supports sparse weight compression to save memory bandwidth. Built-in Winograd convolution support improves computational efficiency for filters of certain sizes. Batch convolution saves additional memory bandwidth by reusing weights when running multiple inferences in parallel.

To avoid duplicate access to system memory, the NVDLA convolution engine has an internal RAM for the storage of weights and input features, called the "convolution buffer." This design greatly improves memory efficiency, instead of sending a request to the system memory controller every time a weight or feature is needed.

Convolution units map to TensorFlow operations, such as tf.nn.conv2d.

Single Data Point Processor

The Single Data Point Processor (SDP) allows linear and nonlinear functions to be applied to a single data point. This is usually used immediately after convolution in CNN systems. SDP has a lookup table to implement nonlinear functions and supports simple biasing and scaling for linear functions. This combination can support the most common activation functions, as well as other element-level operations including ReLU, PReLU, precision scaling, batch normalization, bias addition, or other complex nonlinear functions such as sigmoid or hyperbolic tangent.

SDP maps to TensorFlow operations, including tf.nn.batch_normalization, tf.nn. bias_add, tf.nn.elu, tf.nn.relu, tf.sigmoid, tf.tanh, and others.

Flat Data Processor

The Planar Data Processor (PDP) supports specific spatial operations common in CNN applications. It is configurable at runtime to support different pooling group sizes and supports three pooling functions: max pooling, min pooling, and average pooling.

PDP maps to TensorFlow operations, including tf.nn.avg_pool, tf.nn.max_pool, and tf.nn.pool, among others.

Cross-channel Data Processor

The Cross-channel Data Processor (CDP) is a specialized unit used to implement the local response normalization (LRN) function—a special normalization function in the channel dimension, rather than operating in spatial dimensions.

CDP maps to the tf.nn.local_response_normalization function.

Data Reshaping Engine

The data reshaping engine performs data format transformations (e.g., split or slice, merge, shrink, reshape—transpose). During the inference process of convolutional networks, the data in memory often needs to be reconfigured or reshaped. For example, the "slice" operation can be used to separate different features or spatial regions of an image, and the "reshape-transpose" operation creates output data with a larger dimension than the input dataset, which is common in deconvolutional networks.

The data reshaping engine maps to TensorFlow operations such as tf.nn.conv2d_ transpose, tf.concat, tf.slice, and tf.transpose .

Bridge DMA

Bridge DMA (BDMA) module implements the data copy engine for moving data between system DRAM and a dedicated high-performance memory interface (if present). This is an accelerated path for moving data between these two unconnected memory systems.

Configurability

NVDLA has a range of hardware parameters that can be configured to balance area, power, and performance. Below is a short list of these options.

- **Type of data**. NVDLA natively supports multiple data types in its various functional units; a subset of these can be selected to save area. Selectable data types include binary, INT4 (4-bit integer), INT8, INT16, INT32, FP16 (16-bit floating point), FP32, and FP64.
- **Input image memory format**. NVDLA can support planar images, semi-planar images, or other compressed memory formats. These different modes can be enabled or disabled to save area.
- **Weight compression**. NVDLA has a mechanism to reduce memory bandwidth by sparsely storing convolution weights. This feature can be disabled to save area.
- **Winograd convolution**. The Winograd algorithm is an optimization of certain dimensions of convolution. NVDLA can be built with or without support for it.
- **Batch convolution**. Batch processing is a memory bandwidth saving feature. NVDLA can be built with or without support for it.
- **Convolution buffer size**. The convolution buffer consists of many banks. You can adjust the number of groups (from 2 to 32) and the size of each group (from 4 to 8 KiB). (Multiplying these two numbers determines the total amount of convolution buffer memory that will be instantiated.)
- **MAC array size**. The multiply-accumulate engine is represented in two dimensions. The width ("C" dimension) can be adjusted from 8 to 64, and the depth ("K" dimension) can be adjusted from 4 to 64. (The total number of multiplies created can be determined by multiplying these two.)
- **Second memory interface**. NVDLA can support a second memory interface for high-speed access, or it can be built with just one memory interface.
- **Nonlinear activation function**. To save area, look-up tables supporting nonlinear activation functions (such as sigmoid or tanh) can be removed.
- **Activate engine size**. The number of active outputs generated per cycle can be adjusted between 1 and 16.
- **Bridge DMA engine**. The bridge DMA engine can be removed to save area.
- **Data reshaping engine**. The data reshaping engine can be removed to save area.
- **The present of a pooling engine**. The pooling engine can be removed to save area.
- **The size of a pooling engine**. The pooling engine can be tuned to produce 1–4 outputs per cycle.
- **The present of a local response normalization engine**. The local response normalization engine can be removed to save area.
- **The size of a local response normalization engine**. The local response normalization engine can be tuned to produce 1–4 outputs per cycle.
- **Memory interface bit width**. The memory interface bit width can be adjusted according to the width of the external memory interface, thereby appropriately adjusting the size of the internal buffer.
- **Memory read latency tolerance**. Memory latency is defined as the period from a read request to the return of read data. The tolerance for this delay can be adjusted, which affects the internal delay buffer size of each read DMA engine.

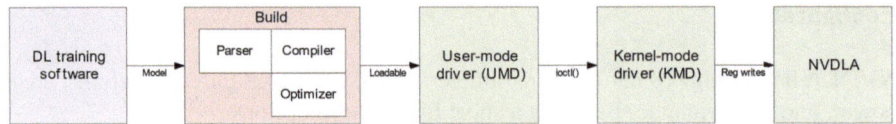

Fig. 7.9 NVDLA system software internal data flow diagram

Software Design

NVDLA has a complete software ecosystem that supports it. Part of this ecosystem includes on-device software stacks. In addition, NVIDIA provides complete training infrastructure to build new models incorporating deep learning and convert existing models into a form usable by NVDLA software. Typically, NVDLA-related software is divided into two groups: *compilation tools* (model transformation) and *runtime environments* (runtime software that loads and executes networks on NVDLA). The general process is shown in Fig. 7.9, and each group is described below.

Compilation Tools: Model Creation and Compilation

Compilation tools include compilers and parsers. The compiler is responsible for creating a series of hardware layers that are optimized for a given NVDLA configuration. Having an optimized network of hardware layers improves performance by reducing model size, load, and runtime. Compilation is a multi-step process that can be broken down into two basic components: parsing and compilation. The parser is relatively simple, its most basic function is to read the pre-trained Caffe model and create an "intermediate representation" of the network to pass to the next step of the compiler. The compiler takes as its input the parsed intermediate representation and the hardware configuration of the NVDLA instance and generates the hardware layer network.

   Understanding the specific hardware configuration of an NVDLA instance is important to enable the compiler to generate the appropriate hardware layer for the available feature set. For example, choose between different convolution operation modes (such as Winograd convolution or basic convolution) or split the convolution operation into multiple smaller mini operations based on the available convolution buffer size. This stage is also responsible for quantizing the model to lower precision, such as 8- or 16-bit integers, or 16-bit floating-point numbers, and allocating memory areas for the weights. The same compiler tool can generate action lists for multiple different NVDLA configurations.

Runtime Environment: Model Inference on the Device

The runtime environment involves running the model on compatible NVDLA hardware. It is divided into two layers:

- **User-mode driver**. **The main** interface provided to user-mode programs. The neural network compiler parses and compiles the network layer by layer and converts it into a file format called NVDLA Loadable. The user-mode runtime driver loads it and submits the inference job to the kernel-mode driver.
- **Kernel-mode driver**. Consists of drivers and firmware that are responsible for scheduling "layer" operations on NVDLA and programming NVDLA registers to configure each functional block.

Runtime execution starts from a network representation of the storage; this storage format is called an "NVDLA Loadable" image. From a loadability perspective, each functional block in an NVDLA implementation is represented by a "layer" in the software. Each layer contains information about its dependencies, input and output tensors in memory, and an operation. Specific configuration for each block in. Layers are linked together through a dependency graph, which is used by kernel-mode drivers to schedule each operation. The NVDLA loadable file format is standardized and implemented across compilers and user-mode drivers.

The user-mode driver has a standard application programming interface (API) for handling loadable images, binding input, and output tensors to memory locations, and running inference. It loads the network into memory as a defined set of data structures and passes them to the kernel-mode driver in a defined implementation. For example, on Linux, this might be an ioctl () that passes data from a user-mode driver to a kernel-mode driver, whereas on a single-process system (the user-mode driver runs in the same environment as the kernel-mode driver), which could be a simple function call.

The main entry point of the kernel-mode driver receives an inference job in memory, selects it from multiple available jobs for scheduling (if on a multi-process system), and submits it to the core engine scheduler. This core engine scheduler is responsible for handling interrupts from NVDLA, scheduling the layers on each individual functional block, and updating all dependencies of that layer based on the completion of the previous layer's tasks. The scheduler uses information from the dependency graph to determine when subsequent layers are ready for scheduling, which allows the compiler to decide the scheduling of layers in an optimized way and avoid performance differences between different implementations of kernel-mode drivers.

Area and Performance Estimates for NVDLA

Table 7.2 presents area and performance estimates for NVDLA configurations optimized for ResNet-50 neural networks. The area figures given are estimates of the combined area including all storage required. In this example, no on-chip SRAM is used. If the available SDRAM bandwidth is low, on-chip SRAM would be beneficial.

Table 7.2 NVDLA area and performance estimates

Number of multiply-accumulate engines	Convolution buffer size (KB)	SDRAM bandwidth (GB/s)	Silicon component area (mm², 28 nm)	Silicon component area (mm², 16 nm)	Int8 ResNet-50 inference performance (frames/s)	Power estimation peak/average (mW, 16 nm)
2048	512	20	5.5	3.3	269	766/291
1024	256	15	3.0	1.8	153	375/143
512	256	10	2.3	1.4	93	210/80
256	256	5	1.7	1.0	46	135/48
128	256	2	1.4	0.84	20	82/31
64	128	1	0.91	0.55	7.3	55/21
32	128	0.5	0.85	0.51	3.6	45/17

The power and performance in the table are for 1-GHz frequency. By adjusting voltage and frequency, the power and performance of a specific configuration can be changed.

It is precisely because of the above characteristics that Jetson Xavier series modules use NVDLA technology, which is suitable for application on a variety of IoT devices of different sizes and complexity, achieving a balance between high performance, low-power consumption, and small area. Taking Jetson Xavier NX as an example, each NVDLA engine of this module generates 4.5 TOPS of computing power.

7.2.3   Jetson Performance

Table 7.3 compares the main technical indicators of Jetson series modules.

To visually compare the scalability between Jetson series members, NVIDIA officially tested the inference of Jetson Nano, Jetson TX2, Jetson Xavier NX, and Jetson AGX Xavier on popular DNN models for image classification, object detection, pose estimation, semantic segmentation, etc. Performance, as shown in Fig. 7.10, these results were run using JetPack and NVIDIA's TensorRT inference accelerator library, which optimizes the real-time performance of networks running in popular machine learning frameworks such as TensorFlow, PyTorch, Caffe, MXNet, etc. training.

In this test, both the NVDLA engine and the GPU run simultaneously at INT8 precision on the Jetson Xavier NX and Jetson AGX Xavier, while the GPU runs at FP16 precision on the Jetson Nano and Jetson TX2.

Table 7.3 Main technical indicators of Jetson series modules

	Jets on NANO	JETSON TX2 series			JETSON XAVIER NX	JETSON AGX XAVIER Series	
		TX2 4 GB	TX2	TX2i		AGX XAVIER 8 GB	AGX XAVIER
AI performance	472 GFLOPS	1.33 TFLOPS		1.26 TFLOPS	21 TOPS	20 TOPS	32 TOPS
GPU	128-core NVIDIA Maxwell™ GPU	256-core NVIDIA Pascal™ GPU			With 48 Tensor Cores	-core NVIDIA Volta™ GPU with 48 Tensor Cores	-core NVIDIA Volta™ GPU with 64 Tensor Cores
CPU	-core ARM® Cortex®-A57 MPCore	Dual-core NVIDIA Denver 1.5 64-bit CPU and quad-core ARM® Cortex®-A57 MPCore processor			6-core NVIDIA Carmel ARM®v8.2 64-bit CPU 6 MB L2 + 4 MB L3	6-core NVIDIA Carmel Arm® v8.2 64-bit CPU 6 MB L2 + 4 MB L3	8-core NVIDIA Carmel Arm® v8.2 64-bit CPU 8 MB L2 + 4 MB L3
Memory	4 GB 64-bit LPDDR4 25.6 GB/s	4 GB 128-bit LPDDR4 51.2 GB/s	8 GB 128-bit LPDDR4 59.7 GB/s	8 GB 128-bit LPDDR4 (supports ECC) 51.2 GB/s	8 GB 128-bit LPDDR4x 51.2 GB/s	8 GB 256-bit LPDDR4x 85.3 GB/s	16 GB 256-bit LPDDR4x 136.5 GB/s
Storage	16 GB eMMC 5.1	16 GB eMMC 5.1	32 GB eMMC 5.1	32 GB eMMC 5.1	16 GB eMMC 5.1	32 GB eMMC 5.1	
Power	5 W/10 W	7.5 W/15 W		10 W/20 W	10 W/15 W	10 W/20 W	10 W/15 W/30 W
Deep learning accelerator	–	–			2 NVDLA engines	2 NVDLA engines	2 NVDLA engines
Visual accelerator	–	–			–	7-way VLIW vision processor	
Mechanical specifications	69.6 mm × 45 mm 260-pin SO-DIMM connector	87 mm × 50 mm 400 pin connector			69.6 mm × 45 mm 260-pin SO-DIMM connector		100 mm × 87 mm 699-pin connector

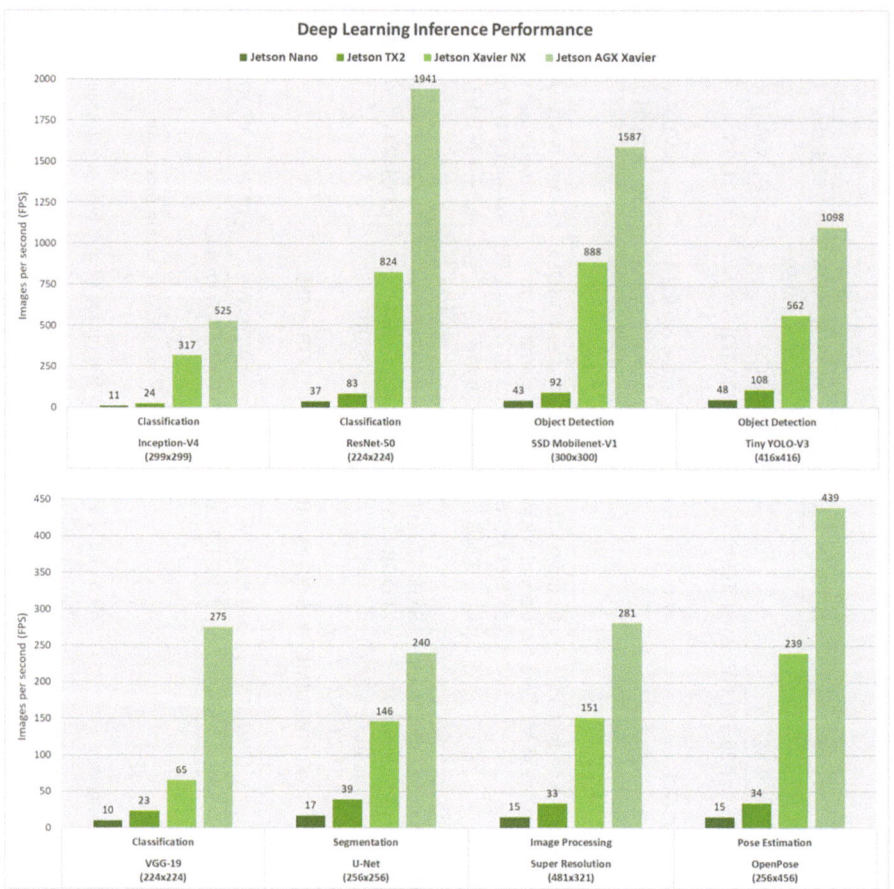

Fig. 7.10 Inference performance of various visual DNN models using TensorRT in the Jetson family. (Source: https://developer.NVIDIA.com/blog/jetson-xavier-nx-the-worlds-smallest-ai-supercomputer/)

7.3   Intel Movidius

Intel Movidius (Intel Corporation, n.d.) is a dedicated AI hardware accelerator with a powerful neural computing engine designed to significantly improve the performance of deep neural networks without compromising their low-power characteristics. Featuring a series of MAC blocks and directly interfaced with intelligent memory structures, this neural computing engine can quickly perform the calculations required for deep inference without encountering the so-called "data wall" bottleneck encountered by other processor designs.

Movidius was once a start-up visual processing chip manufacturer. It was acquired by Intel in 2016 and has launched three generations of products. In order of release time, they are Myriad, Myriad 2, and Myriad X. These chips are specifically designed to process images and video streams. It is positioned as VPU (Vision Processing Units) for its ability to process computer vision.

Myriad is small enough to be packaged in a USB thumb drive and used as a Neural Compute Stick (NCS). The best thing about NCS is that it works with both x86 and ARM devices. It easily plugs into an Intel NUC (Next Unit of Computing) or Raspberry Pi to run inference. It draws power from the host device without requiring an external power source.

Similar to NVIDIA JetPack, Intel has built a software platform to optimize machine learning models for Movidius and Myriad. The Intel OpenVINO Toolkit distribution runs computer vision models trained in the cloud at the edge. OpenVINO, which stands for Open Visual Inference and Neural Network Optimization, is an open-source project that aims to bring consistent inference methods to models running on x86, ARM32, and ARM64 platforms. Existing convolutional neural network (CNN) models can be converted into OpenVINO intermediate representations (IR), which can significantly reduce model size while being optimized for inference.

7.3.1   Movidius Myriad X VPU Chip

The Intel Movidius Myriad X VPU is Intel's third generation and most advanced VPU. It is the first of its kind to feature the Neural Compute Engine—a dedicated hardware accelerator for deep neural network inference. The Neural Compute Engine, combined with 16 powerful SHAVE (Streaming Hybrid Architecture Vector Engine, Streaming Hybrid Architecture Vector Engine) cores and an ultra-high-throughput intelligent storage structure, makes Intel Movidius Myriad X an ideal choice for on-device deep neural network and computer vision applications. Industry leader in programs. Intel's Myriad 8 HD sensors are directly connected to the VPU, as shown in Fig. 7.11 .

The Intel Movidius Myriad X VPU can be programmed through the Myriad Development Kit (MDK), which includes all the necessary development tools, frameworks, and APIs to implement custom vision, imaging, and deep neural network workloads on the chip.

1. Main Features

   (a) Dedicated neural computing engine.
   (b) 16 high-performance SHAVE cores.
   (c) Enhanced ISP with 4K support.
   (d) New visual accelerators including stereoscopic depth of field.
   (e) Native FP16 and 8-bit fixed-point support.
   (f) Quickly port and deploy neural networks in Caffe and TensorFlow formats.
   (g) End-to-end acceleration of many common deep neural networks.
   (h) Industry-leading inference/second/watt performance.

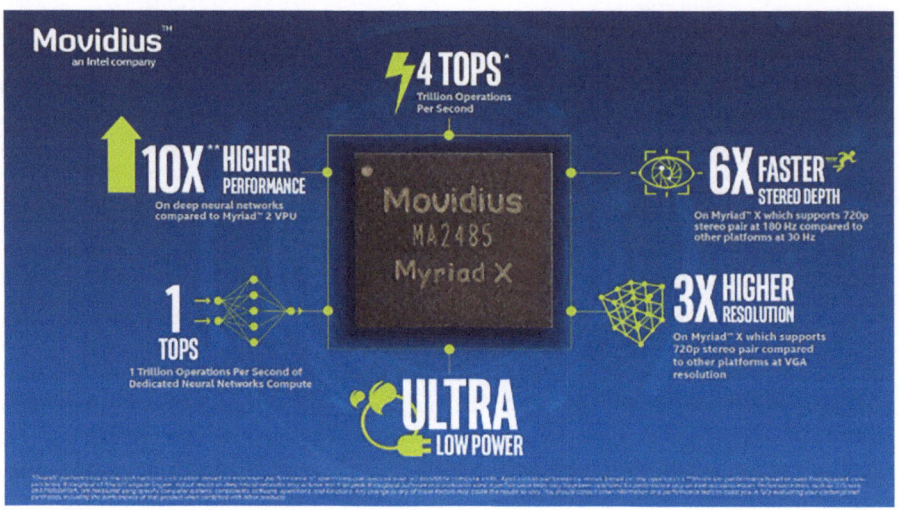

Fig. 7.11 Intel Movidius X chip

2. Neural Computing Engine

Intel Movidius Myriad X is the first VPU to feature the Neural Compute Engine, a dedicated hardware accelerator for running deep neural network applications on the device. With an intelligent storage fabric that interfaces directly with other key components, the Neural Compute Engine can deliver industry-leading performance per watt without the common data flow bottlenecks encountered by other architectures.

3. Programmable SHAVE Core

The Intel Movidius Myriad X chip is manufactured using a 16-nm process and operates at 700 MHz. Features 16 programmable VLIW (Very long instruction word) vector processors, called SHAVE cores, that accelerate neural networks by executing workloads in parallel. Can be used to run traditional computer vision workloads or to enhance the Neural Compute Engine's support for sparse data structures by running custom layer types for CNN applications.

The SHAVE core is a (WikiChip, 2021) VLIW architecture that mixes RISC, DSP, and GPU and has very high-performance numbers. The entire chip is manufactured using a 28-nm process, operates at 600 MHz, and consumes 300 mW of power. For 8-bit integers, its computing power can reach 300 GOPS (or just over 1 TOPS per watt), while for 8-bit vectors, the computing power is 60 GOPS.

The basic concept behind the core of SHAVE is to improve the parallelism of important data. Each SHAVE core is a variable-length very long instruction word (VLLIW) processor that can operate on native 32-bit integer and 128-bit vector values. Each core contains many different types of register files as well as a set of execution units that utilize multiple register files with many ports to process a relatively large set of values at once.

The architecture of the SHAVE core is shown in Fig. 7.12.

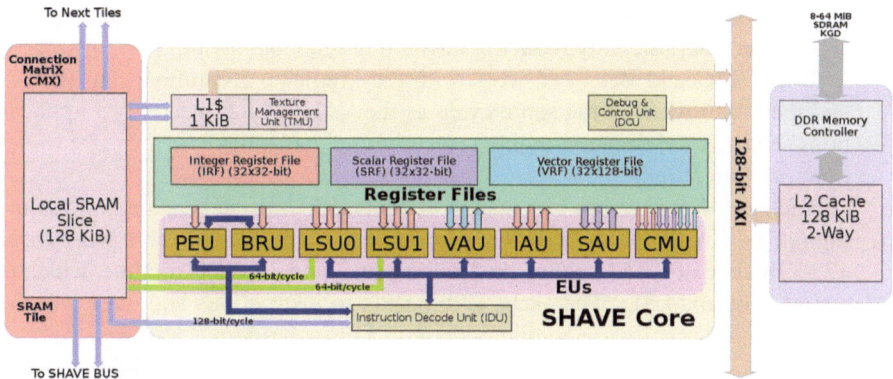

Fig. 7.12 SHAVE kernel architecture

4. Register File

There are three register files: integer register file (IRF), scalar register file (SRF), and vector register file (VRF).

SHAVE's integers register file (IRF) has 17 ports to enable a large number of parallel data operations per cycle. The IRF has 32 entries of 32-bit integers, which are mainly operated by IAU and SAU. In addition to the IRF, there is the Scalar Register File (SRF), which has another set of entries for 32-bit integers. The reason for the second register is to allow SHAVE to do various three-operand operations and generate predicates for the execution unit.

The SHAVE core also has a dedicated vector register file (VRF) containing 32 entries, each 128 bits wide.

5. Execution Unit

The three main arithmetic execution units are the vector arithmetic unit (VAU), the scalar arithmetic unit (SAU), and the integer arithmetic unit (IAU). In addition, each SHAVE core integrates its own texture mapping unit (TMU), allowing powerful bitmap conversions.

The Integer Arithmetic Unit (IAU) executes all arithmetic instructions that operate on 32-bit integers and access the IRF. The Scalar Arithmetic Unit (SAU) is more versatile and can perform all integer (8–32 bit) and floating-point (HP/FP) operations. The Vector Arithmetic Unit (VAU) supports 128-bit vector operations of all integers (8–32 bit) and floating-point (HP/FP) types. Since these operations can be completed in the same cycle, the two can be combined to perform various DSP-like operations, such as matrix transformations.

In addition to these units, there is a comparison move unit (CMU), which is used to generate predicates. This unit can perform operations such as three comparison operations per 16-bit/32-bit/8-bit entry in the vector register file and can perform vector operations that generate predicates for predicates. The CMU efficiently interacts with all register files and can move data between them.

The Branch and Repeat Unit (BRU) is another execution unit capable of continuous iterations with zero overhead. This applies to single or multiple

VLIW words, which can be used as parameterizable finite impulse response (FIR) filters, significantly reducing instruction fetch bandwidth.

Both CMU and BRU can work with the Predicate Execution Unit (PEU), and they both work in the same cycle as the VAU, which means that applications such as pixel decision-making can be completed quickly in the same cycle.

6. Memory Subsystem

Each SHAVE core has two load store units (LSUs), each capable of executing one 64-bit operation from SRAM per cycle. Each SHAVE core has a local 128 KiB slice of SRAM called the Connection Matrix (CMX) module, which is operated by the LSU. The cache is allocated to instructions and data, and the exact amount is software configurable with a granularity of 8 KiB.

Each local cache block is directly linked to its two closest neighbors (presumably the cores located to the west and east), allowing zero-penalty access from these repositories as well. For cache tiles located further away, there does appear to be a slight latency penalty. Movidius notes that most software they've tested does nearly all communication with its neighboring cores, allowing them to take advantage of this.

The entire chip also features a shared 128 KB of L2 cache, and an integrated DDR2 memory controller connected to stacked 8–64 MB SDRAM.

7. Bandwidth

Each register file in the kernel contains many ports. This, coupled with the wide bus, enables very high sustainable throughput at a lower clock frequency (700 MHz).

8. Sparse Data Acceleration

The SHAVE core uses eight 4-bit fields to generate addresses, thereby supporting sparse data operations by loading memory cells.

9. Flexible Image Processing and Encoding

Intel Movidius Myriad X VPU features a fully debug-able ISP pipeline for the most demanding imaging and video applications. The VPU also features hardware-based encoding that delivers up to 4K video resolution, meaning the VPU is a single-chip solution for all imaging, computer vision, and CNN workloads. Up to 8 HD resolution RGB cameras can be connected directly to the Myriad X VPU, each channel of video supports 4K resolution, and overall, supports image signal processing throughput of up to 700 million pixels per second. This brings advanced vision and artificial intelligence applications to devices such as drones, smart cameras, smart homes, security, VR/AR headsets and 360 cameras.

10. Enhanced Vision Accelerator Kit

The Movidius Myriad X VPU adds a new set of visual accelerators, including a new stereo depth module capable of processing dual 720 p signals up to 180 Hz. Leveraging available vision accelerator suites, critical vision workloads can be offloaded to fixed-function hardware, resulting in higher performance at a lower price. Myriad X can leverage more than 20 hardware accelerators to perform tasks such as optical flow and stereo depth without introducing additional computational overhead. For example, the new stereo

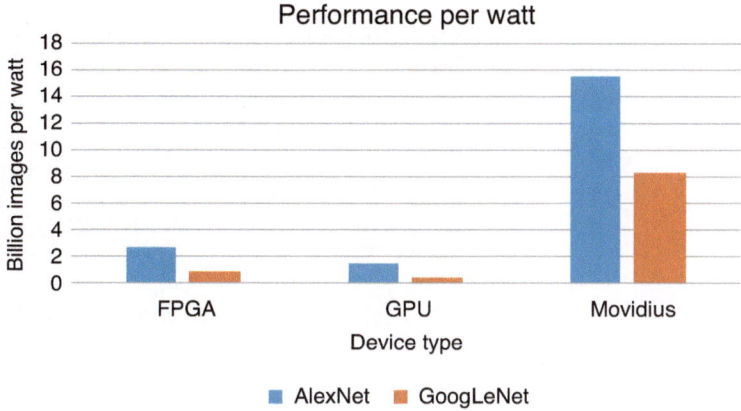

Fig. 7.13 Intel Movidius inference performance. (Source: https://github.com/softserveinc-rnd/fpga-gpu-benchmarking/blob/master/README-MOVIDIUS.md)

depth accelerator can process 6 camera inputs (3 stereo pairs) simultaneously, each at 720 p resolution and 60-Hz frames.

11. Performance

It is precisely because of the adoption of the above technologies that the SHAVE core has excellent performance at ultra-low-power consumption, can provide a total performance of more than 4 trillion operations per second (TOPS), and achieves a calculation of more than 1 TOPS during deep neural network inference performance.

According to information provided by Intel, Movidius' performance per watt has been greatly improved compared to GPU and FPGA acceleration chips. Using the AlexNet model for inference, Movidius can process more than 15 billion images per watt of power, as shown in Fig. 7.13.

7.3.2   Intel Movidius Neural Compute Stick

Intel Movidius Neural Compute Stick 2 (Intel Movidius NCS 2) is a deep learning inference development kit based on Movidius Myriad X. Intel NCS 2 comes in an economical and convenient USB flash stick form. It can be plugged into the USB interface of a PC or embedded device to provide it with additional deep neural network acceleration capabilities, thereby developing and deploying deep learning algorithms at the edge, as shown in Fig. 7.14.

The Neural Compute Stick has low-power consumption and can be used in a variety of embedded devices that can power USB interfaces, such as security cameras, autonomous drones, and industrial machine vision equipment. The convenient USB flash stick form factor makes it easier for developers to create, optimize and deploy advanced computer vision intelligence on a variety of devices at the edge.

Fig. 7.14 Intel Movidius
neural computing stick

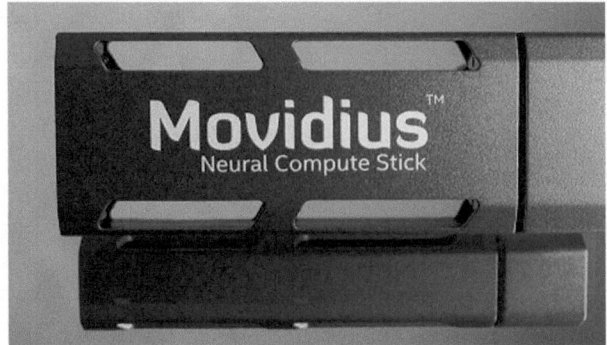

NCS can be used with a trained feed-forward convolutional neural network (CNN). Then, by using a specialized toolkit, the neural network can be analyzed, and the model compiled into an optimized version for embedded deployment.

Here are some of its key features:

- Supports CNN analysis, prototyping and tuning workflows
- All data and power are provided through a single USB Type A port
- Real-time device inference—no cloud connection required
- Run multiple devices on the same platform to scale performance
- Quickly deploy existing CNN models or specially trained networks

Inspired by the subtleties of the human brain's visual system, Intel Movidius is small yet powerful. This allows Movidius TPU and neural computing sticks to become the "eyes" of various embedded devices, processing various situations and things in the real world in real time.

7.4 Google Edge TPU

2016, Google announced that it would add TPU (Tensor Processing Unit) to its cloud platform to accelerate machine learning workloads. Cloud TPU provides the computing power required to train complex machine learning models based on deep neural networks. In a single Cloud TPU V3, 90 TFLOPS of performance can be delivered with 32 GB of high-bandwidth memory (HBM). Customers using Google Cloud Platform can use cloud TPUs from custom virtual machines to balance processor speed, memory, and high-performance storage resources to run compute-intensive workloads.

In 2018, Google announced the launch of Edge TPU (Google LLC., n.d.-a), which is a TPU that can run on the edge (or embedded devices). Edge TPU provides AI computing capabilities in environments where cloud TPU cannot operate by performing inference on trained models at the edge.

7.4.1 Introduction to Google Edge TPU

Edge TPUs are specialized ASICs designed to run AI at the edge. It provides high performance with a small physical and power footprint, allowing for the deployment of high-accuracy AI at the edge, as shown in Fig. 7.15. These specialized chips are used in emerging scenarios such as predictive maintenance, anomaly detection, machine vision, robotics, speech recognition, etc.

A single Edge TPU performs at 4 TOPS (that is, 4 trillion operations per second) using 0.5 W of power per TOPS, or 2 TOPS per watt. How this translates into application performance depends on a variety of factors. Each neural network model has different requirements, and if you are using an embedded device, the overall performance also depends on the host CPU, USB speed, and other system resources.

Table 7.4 compares the time it takes to perform one inference on several popular models on Edge TPU. For comparison, all models run on CPU and Edge TPU are TensorFlow Lite versions.

Among them, the desktop CPU uses 64-bit Intel Xeon Gold 6154 CPU at 3.00 GHz, and the embedded CPU uses quad-core Cortex-A53 @ 1.5 GHz.

Compared to CPU, Edge TPU improves the performance of AI inference on embedded devices by 1–2 orders of magnitude. Taking MobileNet V2 as an example, it can process video streams with a performance of 400 frames per second.

Based on Edge TPU, Corel has launched a series of embedded hardware boards (Google LLC., n.d.-b) that can be used in prototype environments, production environments and sensor applications, as shown in Fig. 7.16.

Taking the simplest accelerator module as an example, it encapsulates an Edge TPU with a three-dimensional size of only 15 mm × 10 mm × 1.5 mm and a weight of only 0.67 g, but it can perform 4 trillion operations per second using 2 W of power, as shown in Fig. 7.17.

In terms of software, Edge TPU does not need to build models from scratch. Developers can convert TensorFlow models into Edge-compatible TensorFlow Lite

Fig. 7.15 Dimensions of Google TPU

Table 7.4 Google Edge TPU inference performance for different models (source: https://coral.ai/docs/edgetpu/benchmarks/)

Model architecture	Desktop CPU	Desktop CPU with Edge TPU + USB accelerator (USB 3.0)	Embedded CPU	Development board with Edge TPU
Unet Mv2 (128 × 128)	27.7	3.3	190.7	5.7
DeepLab V3 (513 × 513)	394	52	1139	241
DenseNet (224 × 224)	380	20	1032	25
Inception v1 (224 × 224)	90	3.4	392	4.1
Inception v4 (299 × 299)	700	85	3157	102
Inception-ResNet V2 (299 × 299)	753	57	2852	69
MobileNet v1 (224 × 224)	53	2.4	164	2.4
MobileNet v2 (224 × 224)	51	2.6	122	2.6
MobileNet v1 SSD (224 × 224)	109	6.5	353	11
MobileNet v2 SSD (224 × 224)	106	7.2	282	14
ResNet-50 V1 (299 × 299)	484	49	1763	56
ResNet-50 V2 (299 × 299)	557	50	1875	59
ResNet-152 V2 (299 × 299)	1823	128	5499	151
SqueezeNet (224 × 224)	55	2.1	232	2
VGG16 (224 × 224)	867	296	4595	343
VGG19 (224 × 224)	1060	308	5538	357
EfficientNet -EdgeTPU-S*	5431	5.1	705	5.5
EfficientNet -EdgeTPU-M*	8469	8.7	1081	10.6
EfficientNet Net-EdgeTPU-L*	22,258	25.3	2717	30.5

* Latency on CPU is high for these models because the TensorFlow Lite runtime is not fully optimized for quantized models on all platforms

models and deploy them on Edge TPU. Google has made a web-based command line tool to perform this conversion. Unlike NVIDIA and Intel edge platforms, Google's Edge TPU cannot run models other than TensorFlow at the edge. But with Google AutoML Vision Edge, the workflow involved in training models in the cloud

Fig. 7.16   Google Edge TPU embedded board

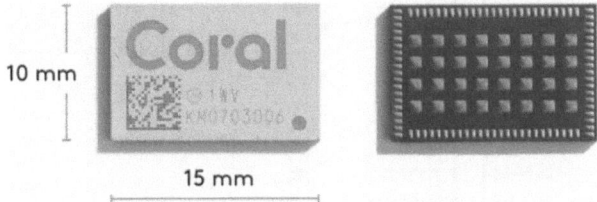

Fig. 7.17   Google Edge TPU module

and deploying them on Edge TPU can be simplified and automated. This is a simple workflow you can use to build a cloud-to-edge pipeline.

7.4.2   How Google Edge TPU works

Google TPU is designed to imitate the working principle of neurons. Here's how Google's Edge TPU hardware is built (Q-engineering, 2021).

Adder

The three main components of a neuron, multipliers, adders, and activation functions must be included in the hardware. First is the adder.

Figure 7.18 is the hardware of a 4-bit adder. A and B are inputs. If the output overflows C4, the carry bit is set. C0 is the carry from the previous stage.

Each basic number gate (AND, OR, and NOT) has its own symbol. They usually consist of two or three transistors. Signals A and B propagate through the circuit and produce the result $A + B$. Changing one of these changes the output almost immediately. This happens very quickly, within a few nanoseconds.

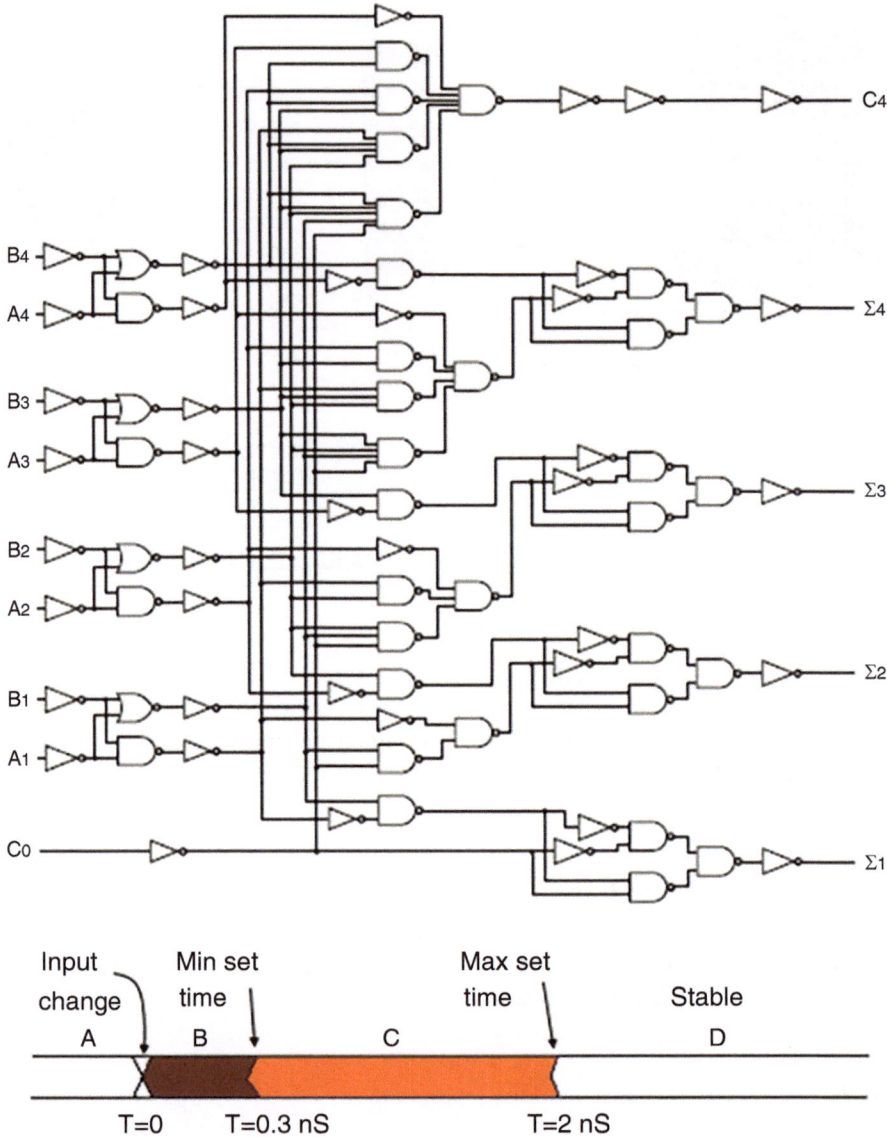

Fig. 7.18 Google Edge TPU adder

The above propagation time depends on the number of digital ports where the output changes. Sometimes two gates simply change their state. And sometimes six gates in a chain must change their outputs. Therefore, the propagation time is not fixed but lies between two limits, the minimum time, and the maximum time, as shown in the text at the bottom of the figure above. All times mentioned are illustrative and have no relation to any device.

Assembly Line

If an adder has a propagation time of 2 ns, the maximum clock rate can be 500 MHz. As mentioned before, a neuron can have hundreds of inputs, all of which must be summed. Designing a chain of adders is not a technical problem. However, the propagation time increases dramatically. The last adder in the chain must wait for all intermediate results before its output stabilizes. For a chain with 250 inputs, each input has a delay of 2 ns, the total time is **500** ns. This results in a very slow 2-MHz clock. The solution to this problem is the pipeline structure. Place a storage element after each adder so that its result remains stable for the next adder, as shown in Fig. 7.19.

The output of the register is updated on the rising edge of the clock signal. That's the only time the output can change. When the clock signal is high, low, or falling, the input cannot manipulate the output and it remains stable. The input must remain stable for a minimum period before the clock rises, and after it rises. Taking 0.1 ns as an example, this register has its own propagation time (0.4 ns). The total propagation time of the new stable signal is now 2.5 ns, which will result in a maximum clock speed of 400 MHz. When an adder is placed after each adder, the situation shown in Fig. 7.20 occurs.

Each color represents a value. Four clock cycles later, the value at the input propagates through the network and appears at the output. Because the registers are updated simultaneously, a new input is accepted every 2.5 ns and propagation starts again. The time it takes to go through the entire pipeline is called latency, which in the diagram above is 10 ns (4 × 2.5 ns). This pipeline technology is the backbone of modern computer technology and can be seen everywhere. Each digital component is pipelined to ensure the required speed.

Mul-Add Cell

Each input signal to a neuron has its own impact on the outcome. Each input is multiplied with a weight value. Just like adders, multiplication can be built with the same basic digital gates. As you can imagine, more gates are now needed to perform multiplication than summation. In the neuron's formula, each input is multiplied by its own weight value. Therefore, a multiply-accumulate unit is introduced, as shown in Fig. 7.21.

Considering the formulation of neurons, it is relatively easy to design neurons with multiple mul-add cells. Figure 7.22 shows the implementation structure of a neuron with three inputs.

An important point to mention is the registers at inputs X1 and X2. They create a delay line and are placed here to synchronize the necessary delays in the mul-add chain. When the second mul-add cell accumulates the output of the first cell (signal A), it is delayed by one clock cycle. Therefore, the multiplication of X1 and its weight also needs to be delayed by one cycle in order to get the results at the same time. The color scheme at the top of the image makes everything clear at a glance.

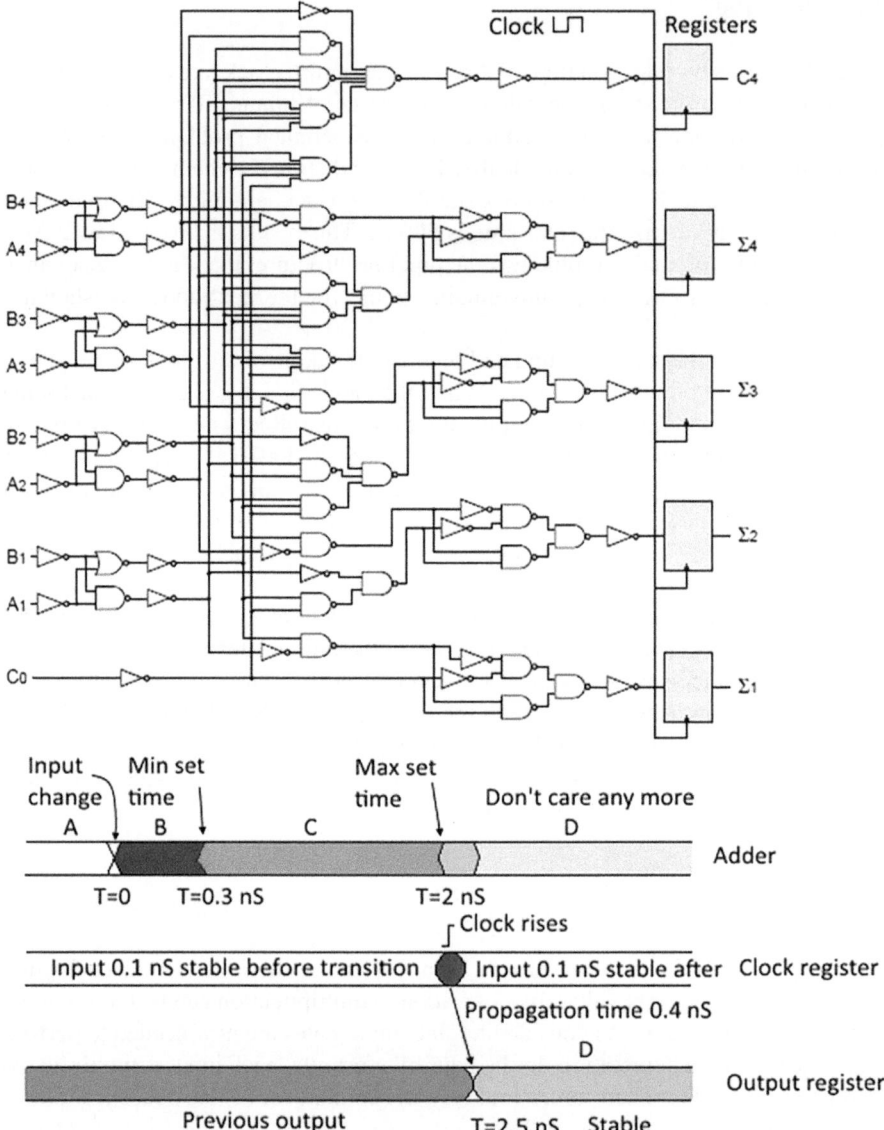

Fig. 7.19 Google Edge TPU pipeline structure

Each color is an element of the input vector [X0, X1, X2]. Looking at the horizontal lines, inputs A and B always have the same color. This means they occur at the same time; they are synchronized. The same applies to inputs C and D.

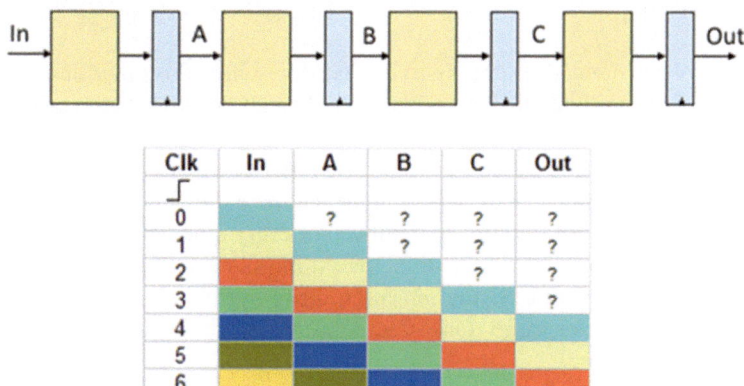

Fig. 7.20 Google Edge TPU pipeline propagation time

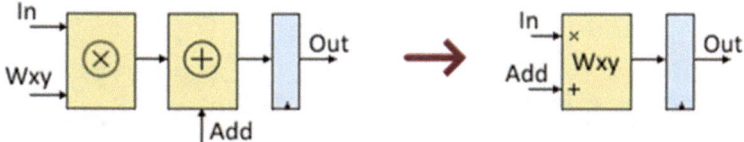

Fig. 7.21 Multiply-add unit

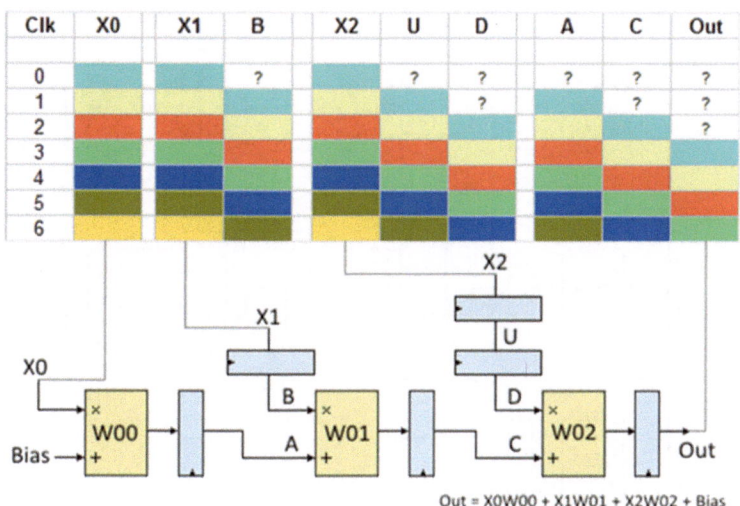

Out = X0W00 + X1W01 + X2W02 + Bias

Fig. 7.22 Implementation structure of a neuron with three inputs

Pulsating Array

Once a neuron is created, it is easy to extend this scheme to other neurons on the same layer. Each individual input is connected to all neurons in the next layer, as shown in Fig. 7.23.

This form of design is called a systolic array. All values are pumped step by step from top to bottom, like a heart pulsating, hence the name. From a timing perspective, the array can accept a complete input vector every clock cycle. So, in the above example, the propagation time is still 2.5 ns. At the same speed, the array generates the complete output vector. If the pulsation array expands in depth or width, the travel time remains the same. Only the delay time will increase. This powerful

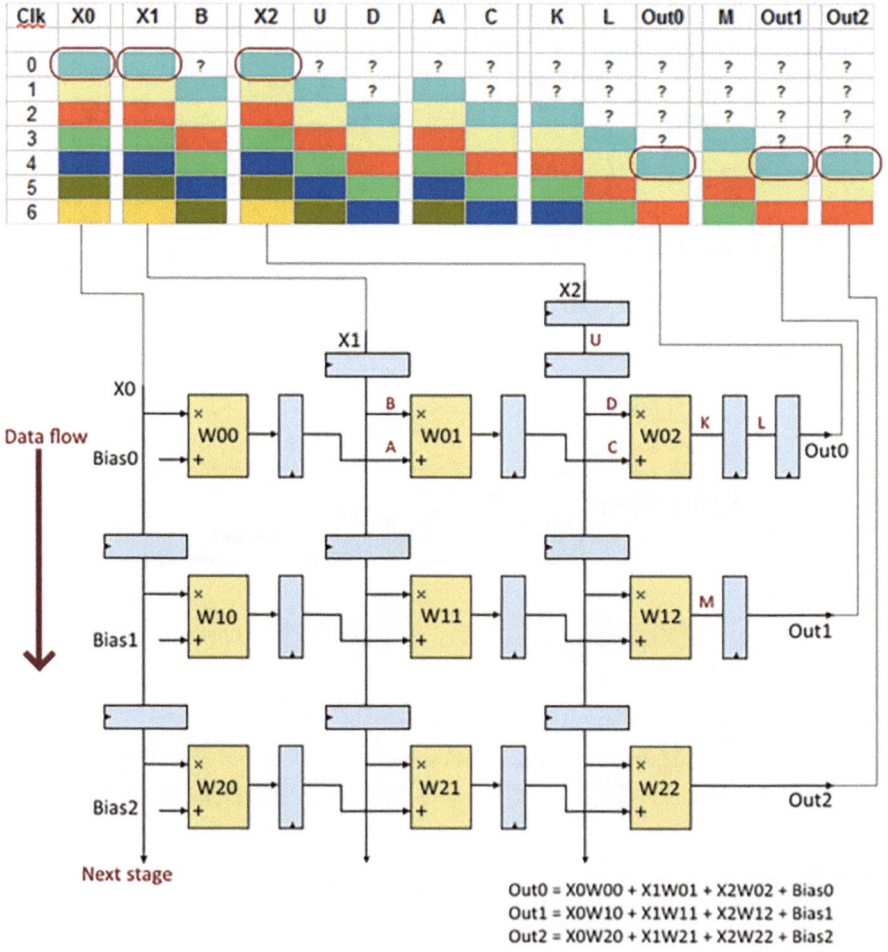

Fig. 7.23 Systolic array

parallel computing capability is why systolic arrays are widely used in neural network hardware.

The size of the systolic array in the Edge TPU depends on the specific model. The first TPU used by Google in its data centers contained 256×256 multiply-add cells. It runs at 700 MHz and can theoretically perform $256 \times 256 \times 700,000,000 = 46$ trillion mul-add operations per second. Or, calculated as a single operation, 92 trillion operations per second (92 TOPS). However, the numbers above are purely theoretical. In practice, several factors can cause performance degradation.

Systolic arrays consist of hardware. This means it has fixed dimensions. Therefore, the input and output vectors also have fixed dimensions. However, the number of neurons per layer is determined by the design and is certainly not constant. If this amount is smaller than the size of the array, you can of course extend the input vector with leading zeros so that the dimensions are equal to the array size. The same technique can be applied to output vectors that are too large.

If the input vector contains more elements than the width of the systolic array, the entire arithmetic calculation must be performed in multiple calls. To do this, the input vector is cut into a series of smaller vectors that match the width of the array. Then process them one by one. Intermediate results are stored in buffers. These values are accumulated after all pulse calls are completed.

Note that not only must the input vector be split into smaller parts, but all weights used in the multiplication must be updated according to the processed part of the formula. This may require large memory transfers. Below is an example of a systolic queue with only four inputs processing a vector of dimension 12.

$$\mathrm{Out}_A = X_0 \cdot W_0^0 + X_1 \cdot W_0^1 + X_2 \cdot W_0^2 + X_3 \cdot W_0^3 \rightarrow \mathrm{Out} = \mathrm{Out}_A + \mathrm{Out}$$
$$\mathrm{Out}_B = X_4 \cdot W_0^4 + X_5 \cdot W_0^5 + X_6 \cdot W_0^6 + X_7 \cdot W_0^7 \rightarrow \mathrm{Out} = \mathrm{Out}_B + \mathrm{Out}$$
$$\mathrm{Out}_C = X_8 \cdot W_0^8 + X_9 \cdot W_0^9 + X_{10} \cdot W_0^{10} + X_{11} \cdot W_0^{11} \rightarrow \mathrm{Out} = \mathrm{Out}_C + \mathrm{Out}$$

Figure 7.24 shows a schematic diagram of the Google TPU, with the output buffer and accumulator at the bottom of the figure.

In the above picture, a buffer is also placed on the input vector side. They act as first-in-first-out buffers or FIFOs. This ensures continuous input to the systolic array.

Activate Unit

Once the output is available, it is sent to the activation unit. This module in the Edge TPU applies an activation function to the output. It's hardwired. In other words, the feature cannot be changed, it works like a ROM. This unit is probably a ReLU function since it is the most used activation function today and is easily implemented in hardware.

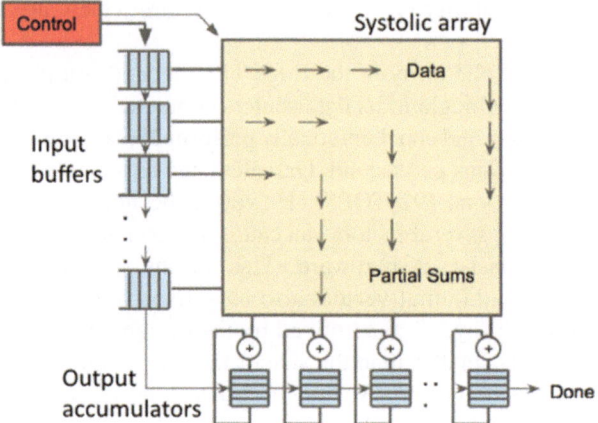

Fig. 7.24 Google Edge TPU

The Difference Between TPU and Edge TPU

Google's Edge TPU is much smaller than the TPUs Google uses in its data centers. It's obvious that such a small device can't possibly have the same functionality as its larger ancestor. The chip layout does not allow this to happen. The Pulse Array also doesn't have the same 256 × 256 dimensions as the original TPU. Google hasn't revealed the specs in Edge so far, but an educated estimate is 64 × 64 clocked at 480 MHz, allowing for 4 TOPS of inference performance.

Memory is also an issue. In the original TPU chip, it accounts for about 29% of the total layout. It is not possible to have the same amount of memory on the Edge TPU chip. In addition, the Edge TPU chip is so energy efficient that it requires no additional cooling. This means that all buffering is almost certainly off-chip. It only implements functions such as input vectors, output vectors, and weights.

Practice

In fact, this is the design philosophy of the Google Edge TPU. Offload all tensor calculations to the TPU as quickly as possible, get the results, populate them to the next layer, and start the loop again from this layer. Work on it layer by layer until done.

To speed up this process, TensorFlow uses a special backend compiler, the Edge TPU compiler. The main task of this Edge TPU compiler is to split TensorFlow Lite files into more suitable TPU transmission packages. If for any reason the TPU cannot process a TensorFlow Lite file or part of it, the CPU will process it. However, if this happens, the application will of course run very slowly.

8-Bit Integer

Another way to speed up calculations is to use 8-bit signed integers instead of floating-point numbers. Floating-point numbers occupy 32 bits in memory, while integers only require 8 bits. Because neural networks are insensitive to numerical precision, it also works well with 8-bit integers. It will still maintain accuracy. This technology not only saves about 75% of memory, but also significantly reduces the number of transistors on the chip.

The conversion algorithm from floating-point numbers to 8-bit signed integers is very simple. First, it determines the maximum and minimum values that the variables in the model will take. Then take the larger of the two and scale it to 127. Assume the minimum value is -1205.8 and the maximum value is 646.8. The larger one is the absolute value of the minimum value 1205.8. So, the number becomes -127. Zero remains zero. The integer value of 646.8 is $127 \times (646.8/1205.8) = 68$. Every number in a TensorFlow model is transformed this way. Not only for Edge TPU, by the way, all TensorFlow Lite models for embedded deep learning are handled this way.

The output of the Edge TPU is a floating-point number. This is because the activation function is embedded in ROM, so storing floating-point numbers is as easy as storing integers.

To adapt to the characteristics of Edge TPU 8-bit integer inference, quantization-aware training should be used in TensorFlow. It simulates the effect of 8-bit signed integers, thus producing a more accurate model. It makes the model more tolerant of low precision variables as well.

In short, Google Edge TPU is a purpose-built ASIC from Google designed to run AI at the edge. It provides high performance with small physical size and energy consumption, enabling the deployment of high-accuracy artificial intelligence on embedded devices.

7.5 XILINX DPU

XILINX is the world's leading provider of complete programmable logic solutions and has a monopoly in the FPGA market. For the inference of deep neural networks, XILINX launched the AI acceleration chip DPU (Deep Learning Processing Unit) (XiLinx, n.d.) from the cloud to the edge.

7.5.1 Function

The XILINX deep learning processing unit is a programmable computing engine optimized for convolutional neural networks. The degree of parallelism used in the engine is a design parameter that can be chosen based on the target device and application. It contains a set of highly optimized instructions and supports most convolutional neural networks such as VGG, ResNet, GoogLeNet, YOLO, SSD, MobileNet, FPN, etc.

The DPU has the following features:

1. An AXI slave interface for accessing configuration and status registers.
2. An AXI master interface for accessing instructions.
3. Supports configurable AXI master interface with 64-bit or 128-bit for accessing data based on target device.
4. Supports single per-channel configuration.
5. Supports optional interrupt request generation.
6. Some highlights of DPU capabilities include:

    (a) Configurable hardware architecture cores include: B512, B800, B1024, B1152, B1600, B2304, B3136, and B4096.
    (b) Up to four isomorphic cores.
    (c) Convolution and deconvolution.
    (d) Depth convolution.
    (e) Max pooling.
    (f) Average pooling.
    (g) ReLU, ReLU6, and Leaky ReLU.
    (h) Concat.
    (i) Sum of elements.
    (j) Expansion.
    (k) Reorganization.
    (l) Fully connected layers.
    (m) Softmax.
    (n) Batch normalization.
    (o) Segmentation.

7.5.2   Architecture

The XILINX Deep Learning Processing Unit (DPU) is a programmable engine optimized for convolutional neural networks. It consists of a high-performance scheduler module, a hybrid computing array module, an instruction fetches unit module and a global memory pool module. The DPU uses a specialized instruction set that can efficiently implement many convolutional neural networks. Some examples of convolutional neural networks that have been deployed include VGG, ResNet, GoogLeNet, YOLO, SSD, MobileNet, and FPN, among others.

As an IP (intellectual property nuclear), DPU can be implemented in the programmable logic (PL) of Zynq-7000 SOC or Zynqurascale + MP SOC device and is directly connected to the processing system (PS). The DPU requires instructions to implement the neural network and access storage locations for input images, temporary data, and output data. Programs also need to run on the application processing unit (APU) to handle interrupts and coordinate data transfers.

Figure 7.25 shows the top-level block diagram of the DPU.

Among them, PE (Processing Engine) is the basic processing unit.

The detailed hardware architecture of the DPU is shown in Fig. 7.26. After startup, the DPU obtains instructions from off-chip memory to control the operation of the computing engine. Instructions are generated by the VitisAI compiler, where they are heavily optimized. On-chip memory is used to buffer input, intermediate and output data to achieve high throughput and efficiency. Data is reused wherever possible to reduce memory bandwidth. Deep pipeline design for compute engines. The processing element (PE) takes full advantage of the fine-grained building blocks in XILINX devices, such as multipliers, adders, and accumulators.

As a programmable device, the DPU IP provides some user-configurable parameters to optimize resource usage and customize different functions. Depending on the amount of programmable logic resources available, different configurations can be selected for the use of DSP Slices, LUT, blockRAM, and UltraRAM. There are other functional options such as channel expansion, average pooling, depthwise convolution, and softmax. Additionally, you can determine the number of DPU cores to be instantiated in a single DPU IP.

Among them, the DPU IP can be configured with various convolution architectures, which are related to the parallelism of the convolution unit. The architecture

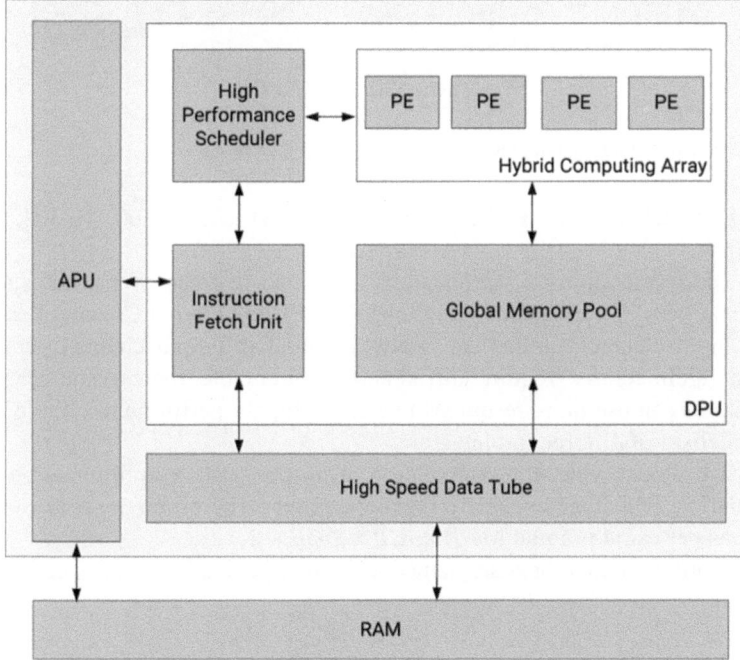

Fig. 7.25 XILINX DPU top-level block diagram

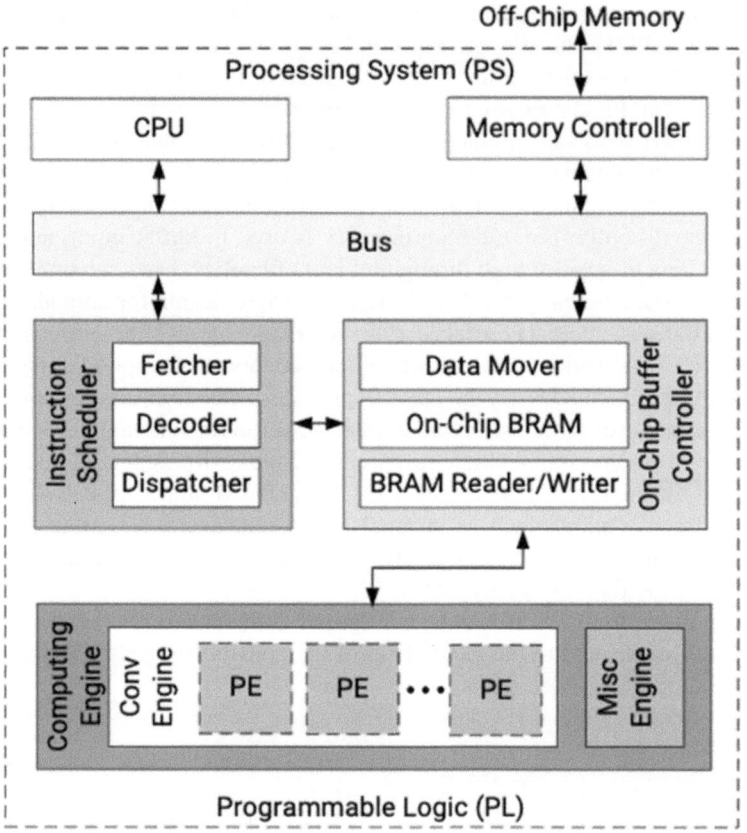

Fig. 7.26 XILINX DPU detailed hardware architecture

of the DPU IP includes B512, B800, B1024, B1152, B1600, B2304, B3136, and B4096.

The DPU convolutional architecture has three degrees of parallelism: Pixel Parallelism (PP), Input Channel Parallelism (ICP), and Output Channel Parallelism (OCP). Input channel parallelism is always equal to output channel parallelism. Different architectures require different programmable logic resources. Larger architectures can use more resources to achieve higher performance. Table 7.5 lists the parallelism of different architectures.

On each clock cycle, the convolution array performs a multiplication and an accumulation, which are considered two operations. Therefore, the peak number of operations per cycle is equal to $PP \times ICP \times OCP \times 2$.

Convolution architectures are named after their peak number of operations.

Table 7.5 DPU convolution architecture configuration table

Convolutional architecture	Pixel parallelism	Input channel parallelism	Output channel parallelism	Peak number of operations (operations/per clock cycle)
B512	4	8	8	512
B800	4	10	10	800
B1024	8	8	8	1024
B1152	8	12	12	1152
B1600	8	10	10	1600
B2304	8	12	12	2304
B3136	8	14	14	3136
B4096	8	16	16	4096

7.5.3   INT8 Optimization

A series of chips with XILINX DPU IP cores are widely used in embedded vision, that is, computer vision algorithms for real-life scenarios are implemented on embedded platforms. Although computer vision algorithms have been significantly improved in recent years, it is a big challenge to transplant such complex and computationally intensive algorithms to embedded platforms while reducing power consumption.

Optimizations of INT8 operations for deep learning also apply directly to traditional computer vision. These algorithms generally work on 8-bit to 16-bit integer expressions. OpenVX is a recently proposed computer vision standard that specifies the use of INT8 expressions per channel (XiLinx, 2017). Most computer vision applications require some degree of filtering, and filtering can be broken down into a set of dot product operations. XILINX's SIMD (Single Instruction Multiple Data) computing mode provides acceleration capabilities for operations involved in visual algorithms.

INT8 calculations inherently take advantage of 27-bit bandwidth. In traditional applications, pre-adders are generally used for efficient implementation $(A + B) \times C$ type operation, but this type of operation is not common in deep learning and computer vision applications. Will $(A + B) \times C$. *The result of C* is decomposed into $A \times C$ and $B \times C$ and then accumulated in independent data streams, making it suitable for the requirements of typical deep learning and computer vision calculations. to INT8 MACC Operationally speaking, we have 18×27 Bit multipliers are dominant. At least one of the inputs to the multiplier must be a minimum of 24 bits, and the carry accumulator must be 32 bits wide to allow 2 INT8 MACC operations to be performed simultaneously on 1 DSP48E2 Slice. The 27-bit input available with 48-bit accumulators are combined to achieve 1.75 times solution performance improvement (1.75:1 ratio of DSP multiplier to INT8 MACC).

In the neural network layer, the main calculation operation for calculating each of the m neuron outputs includes multiplying all n input samples by the corresponding kernel weights and accumulating the results, and then applying the activation function.

If the precision of the sum is limited to INT8, the sum of the products is the first in the parallel MACC introduced in the INT8 optimization method. The second sum of products uses the same inputs but with different kernel weights.

Using the INT8 optimization method to shift the weights 18 bits to the left, each DSP48E2 Slice derives a partial and independent part of the final output value. The accumulator used for each DSP48E2 Slice is 48 bits wide and linked to the next Slice. To avoid saturation caused by weight shifts affecting the calculation, the number of linked modules is limited to 7, that is, $2n$ MACCs and n DSP Slices are used for a total of n input samples. Typical deep neural networks have hundreds to thousands of input samples per layer. But after completing the accumulation of seven terms, the low-order terms of the 48-bit accumulator may be saturated, so an additional DSP48E2 Slice is required for every 7-term sum. This is equivalent to 7 DSP48E2 Slices and 14 MACCs per DSP48E2 Slice, plus one DSP48E2 Slice to prevent oversaturation, resulting in a 1.75× throughput improvement. In a convolutional neural network (CNN), the convolutional layer generally mainly uses the same set of weights, thus forming $A \times W$ and $B \times W$ type parallel MACC operations. Therefore, in addition to input sharing, INT8 optimization can also be used for weight sharing, as shown in Fig. 7.27.

In summary, the XILINX DPU chip can process 2 parallel INT8 MACC operations while sharing the same core weight, thereby optimizing INT8 deep learning and computer vision operations and achieving a 1.75× performance improvement. Used in FPGAs such as Zynq UltraScale + MPSoC, it provides acceleration performance for deep neural network inference on embedded devices.

7.5.4   Performance

Different devices have different DPU IP configurations, different clock frequencies, and different performance. Table 7.6 lists the performance of some devices using DPU IP.

Theoretical peak performance = DPU peak number of operations × Clock frequency.

Among them, ZU9 is configured with 3 B4096 DPU IPs, with a peak performance of 4.1 TOPS.

The inference performance of several models on the XILINX chip is provided below for reference. The results given in Table 7.7 were measured on a XILINX ZCU102 board with 3 B4096 cores and 16 threads running at 287 MHz.

The accuracy rate is the value obtained using 8-bit quantization.

Such performance is sufficient to run some common deep neural network models in real time.

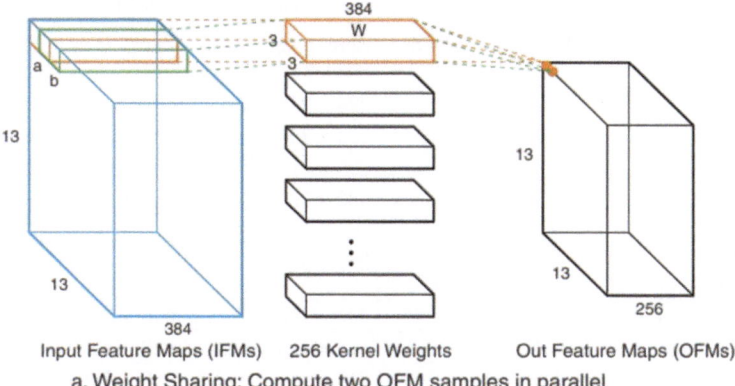

a. Weight Sharing: Compute two OFM samples in parallel

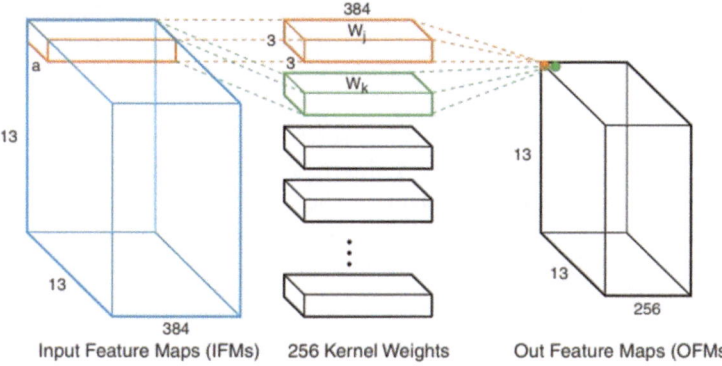

b. Input Sharing: Compute two OFMs in parallel

Fig. 7.27 Multiplication and addition operations in weight sharing and input sharing

Table 7.6 Performance of devices using DPU IP

Device	DPU configuration	Clock frequency (MHz)	Theoretical peak performance (GOPS)
Z7020	B1152x1	200	230
ZU2	B1152x1	370	426
ZU3	B2304x1	370	852
ZU5	B4096x1	350	1400
ZU7EV	B4096x2	330	2700
ZU9	B4096x3	333	4100

7.6   ARM Ethos NPU

ARM is the world's leading embedded CPU manufacturer and has a monopoly in
the mobile phone CPU market. The ARM AI Platform (ARM Limited, n.d.) is a
complete set of heterogeneous computing platforms including ARM Cortex CPUs,

Table 7.7 Performance of common network models running on XILINX ZCU102

Network model	Workload (GOPS/ images)	Enter image resolution	Accuracy (DPU)	Frame rate (FPS)
Inception-v1	3.2	224 × 224	Top-1: 0.6954	452.4
ResNet50	7.7	224 × 224	Top-1: 0.7338	163.4
MobileNet_v2	0.6	299 × 299	Top-1: 0.6352	587.2
SSD_ADAS_ VEHICLE	6.3	480 × 360	mAP: 0.4190	306.2
SSD_ADAS_ PEDESTRIAN	5.9	640 × 360	mAP: 0.5850	279.2
SSD_MobileNet_v2	6.6	480 × 360	mAP: 0.2940	124.7
YOLO-V3-VOC	65.4	416 × 416	mAP: 0.8153	43.6
YOLO-V3_ADAS	5.5	512 × 256	mAP: 0.5301	239.7

Mali GPUs, Ethos NPUs, and microNPUs that deliver advanced machine learning use cases. Coupled with its supporting development environment, ARM has expanded its scope from smartphone CPU design and applications to AI applications for various edge and IoT endpoints.

First launched in late 2019, Ethos is a series of synthesizable neural processor IPs designed by ARM. As part of the Trillium project, the Ethos lineup represents a range of NPUs (Neural Processing Units). The basic microarchitecture of all Ethos NPUs is the MLP (Machine Learning Processor), which can be scaled according to the configuration of SRAM size and the number of compute engines.

7.6.1 ARM Machine Learning Processor

The Machine Learning Processor (MLP) (WikiChip, 2020) is ARM's neural processor microarchitecture. The architecture itself is a base design designed to provide higher performance efficiency for neural network workloads than existing CPUs (Cortex) and GPUs (Mali), with a focus on optimization of CNNs and RNNs. In terms of power consumption, MLP targets a wide range from very low-power embedded IoT applications to complex mobile and network SoCs. Similarly, in terms of performance, MLP uses a configurable expansion design, which can achieve computing capabilities from 1 to 10 TOPS depending on the specific configuration.

MLP is fully statically scheduled by the compiler, which takes a given neural network and maps it to a command stream. This tool chain also has many other optimizations ahead of time, including compression (weights and feature maps are loaded into memory and SRAM libraries) and slicing. The command stream includes necessary DMA operations, such as block fetch operations and accompanying calculation operations. At a higher level, the MLP itself contains a DMA engine, a network control unit (NCU), and a configurable number of compute engines (CE). During operation, the main processor loads a stream of commands onto the control unit, which parses the stream and performs operations by controlling various functional blocks. The DMA engine can communicate with external memory while understanding the various supported neural network layouts, allowing it to handle strides and other predictable neural network memory operations to obtain data in advance for calculations. CE is the main component of the system, and four CEs can form a Quad. CE performs matrix operations and neural network calculations. This architectural design relies on careful co-design with the compiler, resulting in more simplified hardware while achieving more deterministic performance characteristics, as shown in Fig. 7.28.

Calculation Engine

MLP uses a scalable design using a varying number of Compute Engines (CE), which are grouped into Quads. The simplest MLP only requires 1 quad and 1 calculation engine, while the complex MLP has up to 4 Quads and 16 calculation engines. The calculation engine is the main working component of MLP. The three main components in the computing engine are the SRAM library, the MAC Compute Engine (MCE), and the Programmable Layer Engine (PLE). MCE is designed for efficient matrix multiplication, it performs multiply-accumulate operations efficiently on 8-bit integers, while PLE is a more flexible programmable processor designed to enable novel AI functions. Vector operations are supported, and more complex or less common operations can be implemented, as shown in Fig. 7.29.

Convolution Optimization

Each CE comes with a piece of SRAM. All CEs in an MLP are the same size and are configurable in size from 64 to 256 KB. Since the entire execution is statically scheduled, at compile time, the compiler divides the SRAM into multiple parts,

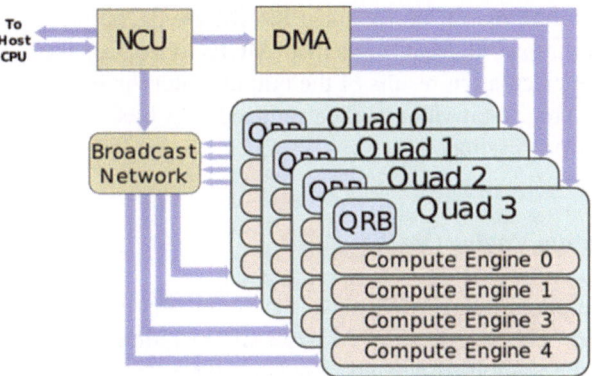

Fig. 7.28 ARM Machine Learning Processor (MLP) architecture

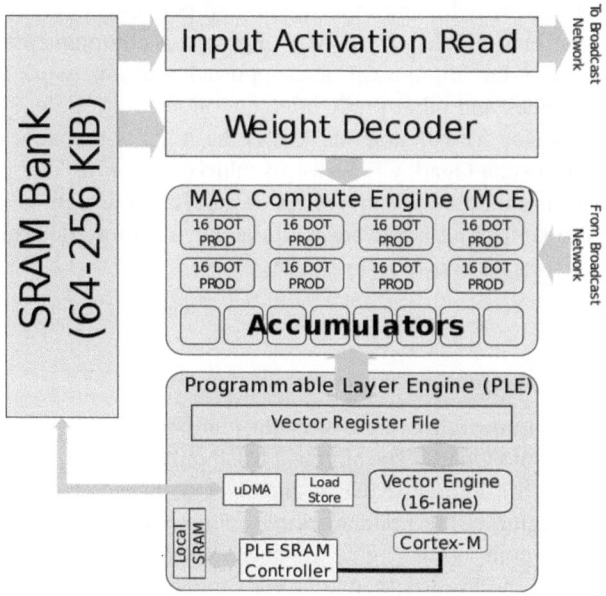

Fig. 7.29 The structure of the computing engine (CE) in MLP

including input feature maps (IFM), compressed form of model weights, and output feature maps.

At runtime, each compute engine is designed to work on a different output feature map, interleaved between compute engines. Input feature maps are interleaved across all SRAM banks. The corresponding weights for each different output feature map are in the SRAM of the CE that processes it. During execution, each compute engine reads a 2D slice (patch) of the input feature map from its local SRAM and sends it to the broadcast network. The broadcast network unit takes these 2D slices and assembles a 3D input activation block (tensor), which is then broadcast to all MCEs in the MLP. As the broadcast continues, the compression weights for a given output feature map from the local SRAM are decompressed and sent to the MAC unit. This allows all MAC units in all MCEs to work on the same input activation block. The calculation results of the output feature map are collected in 32-bit accumulators. The 32-bit value is then reduced to 8 bits and sent by the PLE for additional post-processing.

MAC Computing Engine (MCE)

The MAC Compute Engine (MCE) is designed for very efficient matrix multiplication operations. The weights for each set of operations come from the weight decoder, and the combined activation tensor comes from the broadcast network.

ARM has designed a POP-IP version of MCE that is optimized for the 16- and 7-nm nodes with a custom physical layout design.

Each MCE contains 8 MAC units, each of which can perform 16 8-bit dot product operations. In other words, up to 128 Int8 values can be multiplied by another 128 Int8 values each cycle and accumulated into a 32-bit accumulator. This brings the theoretical peak computing power of each MCE to 256 OPS/cycle. Since each MLP has up to 16 computing engines, the peak computing power of the MLP is 4096 OPS/cycle. Since the target frequency is about 1 GHz, this means that the MLP's computing power can reach up to 4.1 TOPS. Actual performance depends on MCE utilization, which depends on its parameters. MCE has some additional logic to optimize around sparsity, allowing data path gating to zero.

Note that although MCEs are primarily designed for 8-bit operations, they can support 16-bit operations. When working at 16-bit, they run at 1/4 the throughput. In other words, each MCE can perform 64 16-bit OPS/cycle, and all 16 computing units can perform 1024 OPS/cycle. At a frequency of 1 GHz, the Int16 computing power can reach up to 1.024 TOPS.

Finally, the final 32-bit value is reduced to 8 bits before being sent to the PLE for further post-processing.

Programmable Layer Engine (PLE)

The 8-bit results of the MCE are input into the PLE vector register file. The Programmable Layer Engine (PLE) is designed to perform post-processing and enable flexible custom functionality. Once the values arrive in the register file, the CPU starts and operates the vector engine to perform the appropriate operations on the register file. PLE houses an ARM Cortex-M CPU with a 16-channel vector engine that supports vector and neural network extensions, allowing it to perform non-convolutional operations more efficiently.

After processing, the result is returned to the main SRAM bank via DMA. It's worth noting that the data doesn't only come from MCE. For various operations, it is entirely possible for the PLE to obtain data directly from the main SRAM bank to operate on certain data.

7.6.2   Ethos-N Series

The Ethos-N series was launched in October 2019. They are fully independent NPUs that can be integrated into any SoC like the Cortex series. These NPUs have a preconfigured number of computing engines (CE) and quads (quadruples, composed of four CEs). Their performance can reach 1–4 TOPS according to different configurations, and the corresponding energy consumption is 250 mW to 1.5 W. Additionally, the size of the SRAM bank is configurable. Multiple instances of the IP can be interconnected and combined using ARM's CCN-500 or CMN-600

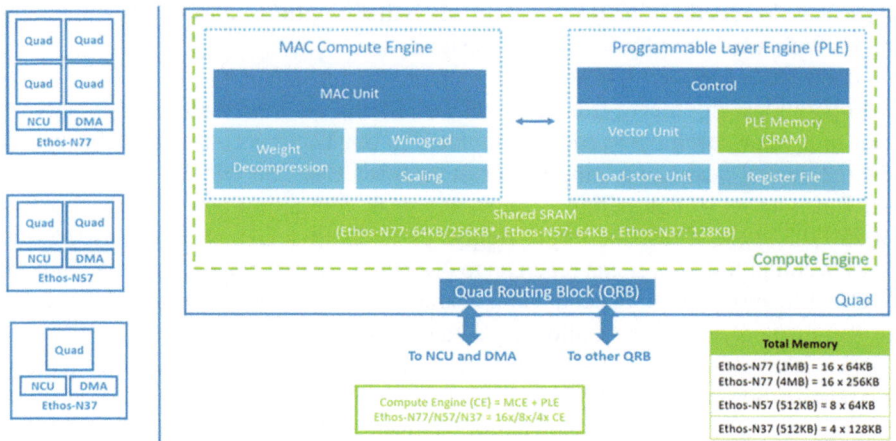

Fig. 7.30 Compute Engine (CE) configuration of Ethos N37, 57, 77

networks to scale to higher performance. For example, 8 Ethos-N77s can be integrated onto a CCN-500 to achieve 32.8 TOPS of processing power, and even higher performance can be achieved using the CMN-600.

Figure 7.30 is the microarchitecture block diagram of the NPU computing engine, and the left side is the configuration diagram of the Ethos-N77, N57, and N37 chips.

Based on the above NPU series, ARM launched Ethos N78, which is a multi-core architecture with 8 NPUs in a cluster and 64 NPUs in the entire mesh, resulting in a considerable increase in performance and efficiency. The new design is significantly higher than the largest Ethos-N77 configuration, now capable of delivering 2× peak performance at up to 10 TOPS of raw compute throughput. ARM has improved the design of the NPU and adopted various new compression technologies to increase the external memory bandwidth of each inference by 40%, thereby improving the power consumption efficiency of the Ethos N78. The Ethos N78's strength is the IP's ability to scale performance across different configuration options. This IP has 4 different performance points or 4 different engine configurations, from the smallest 1 TOPS configuration to 2 TOPS, 5 TOPS, and the largest 10 TOPS configuration, corresponding to MAC configurations of 512, 1024, 2048, and 4096 units, respectively.

In terms of architecture, the biggest improvement of Ethos N78 is the way it processes data in the engine. The new compression method can not only completely compress data outside the NPU and improve DRAM bandwidth, but also move data within the NPU itself, improving performance and power efficiency. Thanks to higher performance density and power efficiency, the average performance improvement from generation to generation is 25%, coupled with double the peak performance configuration, which means it may represent a significant improvement in IoT device performance.

Table 7.8 lists the configuration and performance of Ethos N series NPUs.

Table 7.8 Configuration and performance of Ethos N series NPU

| NPU | Configuration | | SRAM | | | Calculate ability | | |
	Quads	CEs	Bank (KiB)	Total (KiB)	MACs	OPS/ per clock cycle	Int8 performance (GOPS)	Int16 performance (GOPS)
Ethos-N37	1	4	128	512	512	1024	1024	256
Ethos-N57	2	8	64	512	1024	2048	2048	512
Ethos-N77	4	16	64–256	1024–4096	2048	4096	4096	1024
Ethos-N78				384–4096	4096, 2048, 1024, 512		10,240, 5120, 2048, 1024	

7.6.3 Ethos-U Series

The Ethos-U series was launched in early 2020. This series targets deeply embedded AI applications. The microNPUs in this series are not full NPUs like the Ethos-N series. Instead, they feature a slimmer design. The Ethos-U series is designed to work closely with companion Cortex-M processors such as the Cortex-M55. Conceptually, the U series can be viewed as a single Compute Engine (CE) design, with the PLE (Program Layer Engine) module removed and relying instead on the use of the accompanying Cortex-M core for additional processing. Due to power and area constraints, there is only a small amount of built-in SRAM, and the dedicated SRAM bank is removed and instead relies on the shared SoC's storage to save weights and activation values.

Ethos-U has very low-power consumption, with a 90% reduction in machine learning workloads compared to the previous generation Cortex-M. And take up very little memory, and with compression, the model size can be reduced by up to 70%, allowing larger networks to be executed and faster to execute. Offline compilation and optimization techniques for neural networks, operator and network layer fusion, and network layer reordering are performed, further improving performance, and reducing system memory requirements by up to 90%.

With the above features, the Ethos-U series can provide higher performance and lower power consumption, allowing heavy neural network operations to run directly on the micro NPU, such as convolution, LSTM, RNN, pooling, activation function, and original elementwise functions, etc. This supports a variety of popular neural networks, including CNN and RNN, for audio processing, speech recognition, image classification, object detection, etc. Build low-cost, efficient AI solutions in embedded devices.

The Ethos-U series are given in Table 7.9.

The Ethos-U series offers scalable performance and memory interfaces that can be integrated into low-power Cortex-M SoCs as well as high-performance Cortex-A, Cortex-R, and Neoverse SoCs.

Table 7.9 Configuration and performance of Ethos U series

	Configuration		Calculate ability	
NPU	Built-in SRAM (KiB)	MACs	OPS/per clock cycle	Int8 performance (GOPS)
Ethos-U55	18–50	32–256	64–512	25.6–204.8 (@ 100–400 MHz) 64–512 (@ 1 GHz)
Ethos-U	55–104	256–512	512–1024	51.2–409.6 (@ 100–400 MHz) 512–1024 (@ 1 GHz)

Ethos series NPUs all use the Winograd convolution algorithm, and the MAC utilization rate is as high as 90%. Compared with standard convolution, the peak performance can be improved to 2.25 times.

The Ethos series NPUs all use a common tool chain between ARM Cortex and Ethos-U processors, using ARM Endpoint AI Solutions develops, deploys, and debugs AI applications.

Ethos-U55 or Ethos-U65 combined with Cortex-M can provide gesture detection, fingerprint recognition, voice recognition and other capabilities, while the Ethos-N series can complete applications that require higher computing power such as target classification and real-time recognition.

7.7 Qualcomm Hexagon DSP

Hexagon is the trade name for a series of digital signal processor (DSP) products developed by Qualcomm. Also known as QDSP6, representing the "sixth generation digital signal processor," Hexagon architecture is engineered by Qualcomm to provide high performance with low-power consumption across various applications.

Each iteration of Hexagon features its own instruction set and microarchitecture, with these two components intricately intertwined.

Hexagon finds application in Qualcomm Snapdragon chips, employed in smartphones, automobiles, wearable devices, and other mobile gadgets, as well as in components of cellular networks.

To adapt to the increasing demand for machine learning applications, Qualcomm is enhancing Hexagon by integrating matrix multiplication capabilities (clamchowder, 2023).

7.7.1 High Level

Hexagon is a VLIW (very long instruction word) processor characterized by its in-order execution and four-wide design, specifically tailored for specialized signal processing tasks. Utilizing simultaneous multithreading (SMT), Hexagon

Qualcomm Hexagon v73 + Tensor

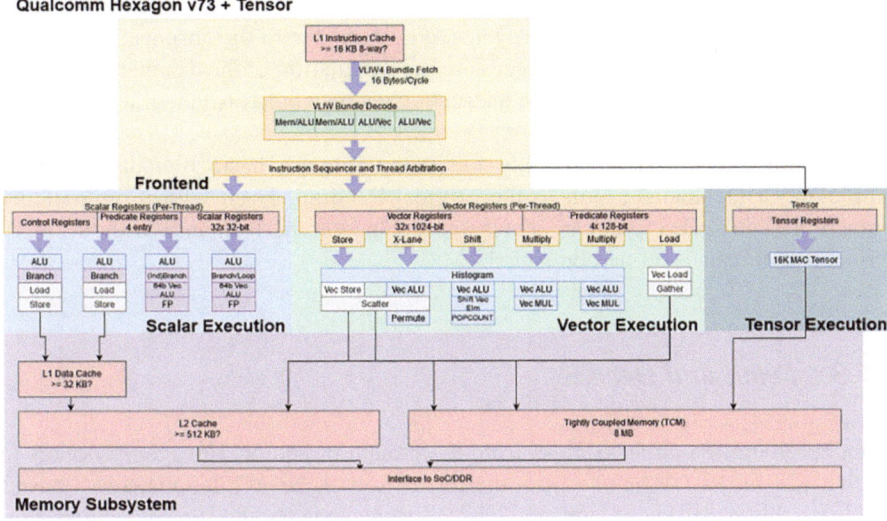

Fig. 7.31 A schematic overview of Hexagon incorporating details extracted from the v73 Programmer's Reference Manuals

efficiently leverages thread-level parallelism to mitigate latency. Its architecture employs a coprocessor model for vector and tensor units, enabling remarkable per-clock throughput capacity. It is shown in Fig. 7.31

Hexagon offers support for virtual memory and caching akin to a traditional CPU, allowing it to execute compiled C code seamlessly. However, it adopts a distinctive execution paradigm where instructions are initially committed before execution. While this approach facilitates deep execution pipelines, it comes with a drawback: certain exceptions can only be identified post-commit, leading to imprecise handling.

Within the Snapdragon 8 Gen 2, Hexagon boasts a 6-way SMT configuration. This positioning places Qualcomm's DSP in a middle ground, falling between GPUs equipped with 12/16 threads per SMSP/SIMD and CPUs, which typically limit themselves to two-way SMT. Each Hexagon thread is initially provided with a scalar context, although it must specifically request access to utilize the vector or tensor coprocessors.

7.7.2 Frontend

Branch prediction is typically a feature found in high-performance CPUs. Nevertheless, Qualcomm has opted to integrate a branch predictor into Hexagon, as indicated by performance monitoring events tracking BTB (Branch Target Buffer) and return stack hits. Descriptions of these events imply that the majority of branches can be managed with a latency of two cycles, occasionally extending to three cycles. When the latency extends to three cycles, it's likely due to the absence of a cached target for the branch in the BTB, necessitating calculation of the destination address by the decoders. Consequently, the instruction cache would also incur a three-cycle latency.

Direction prediction is probably facilitated through basic bimodal counters, employing a mechanism akin to the original Pentium. Even with a rudimentary branch predictor, Hexagon can achieve commendable performance without heavy reliance on thread-level parallelism.

7.7.3   Fetch and Decode

After receiving the target address from the branch predictor, Hexagon proceeds to fetch a 128-bit VLIW bundle from the instruction cache. While Qualcomm hasn't disclosed the exact size of Hexagon's instruction cache for the current version, previous iterations were equipped with 16 KB. It wouldn't be unexpected if Hexagon retained the same instruction cache size, as throughput-centric applications typically exhibit modest instruction footprints.

Haxagon V73 processor architecture is shown in Fig. 7.32

Each VLIW bundle accommodates up to four instructions, prompting contemplation on Qualcomm's choice of the name "Hexagon." "Qualcomm Quadrilateral" might have seemed a more fitting and appealing option. VLIW bundles facilitate superscalar execution with streamlined hardware. Decoders are cost-effective because specific VLIW positions house only a subset of instructions. There's no need for hardware read-after-write or write-after-write hazard resolution since instructions packed into a VLIW bundle must be independent and cannot write to the same destination register. Additionally, execution pipe selection logic is simplified because each VLIW bundle position corresponds to an execution pipe.

Following fetching and decoding, instructions are dispatched to the appropriate unit or coprocessor for execution.

7.7.4   Scalar Integer Execution

Hexagon features a 32-bit scalar unit with 32 registers per thread context, boasting significant capability compared to AMD's GCN or RDNA scalar units. Unlike these, which primarily handle control flow and address generation, Hexagon's scalar unit excels in lightweight DSP tasks independently. With VLIW packing enabling up to four instructions per cycle, each instruction carries out substantial work. Specialized instructions cater to tasks like sum of absolute differences (SAD), bit-field manipulation, and context-adaptive binary arithmetic coding (CABAC) for

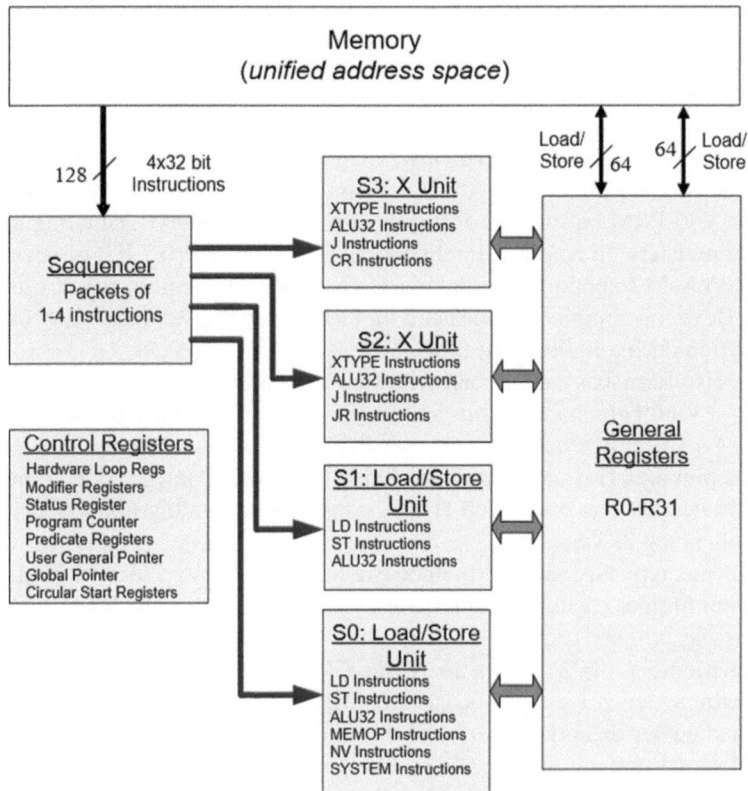

Fig. 7.32 Haxagon V73 processor architecture, From the Hexagon V73 Programmer's Reference Manual

H.264 decode. Additionally, the scalar unit can execute vector operations, allowing byte or halfword elements to be packed into a 32-bit register.

Control registers facilitate hardware looping, enhancing instruction density and register utilization. Circular buffer handling and loop support are bolstered by dedicated registers, freeing general-purpose registers for other data. Although hardware looping aids the branch predictor, mispredictions may occur with short loops. Circular buffers receive hardware support via specific registers, enabling their usage without bounds checking for each pointer increment. A "global pointer" register facilitates access to global or static data using relative addressing mode, alleviating register pressure.

Scalar side memory accesses are managed by an unspecified size L1 cache, historically a 32 KB L1D. With only two 64-bit ports dedicated to the scalar unit, the L1 data cache ensures efficient operation. L1 misses are directed to an L2 cache, with historical precedents suggesting sizes ranging from 256 to 512 KB. Although specifics on the current L2 cache size are not provided, the v73 PRM mentions a 32-entry L2 scoreboard, indicating its capacity to handle pending requests.

7.7.5 Vector Execution (HVX)

Hexagon Vector Extensions (HVX) beefs up processing power for demanding DSP tasks, furnishing 32 1024-bit vector registers and a suite of execution pipes aligning with VLIW bundle positions. Typically, Hexagon DSPs offer fewer vector contexts than scalar ones, necessitating threads to request HVX access. For instance, the Hexagon V73 PRM outlines an example with 4 vector contexts, equating to 16 KB of vector registers. In contrast, Intel's Skylake-X allocates 10.7 KB of vector registers for AVX-512 support. By making HVX capability request-based rather than default, Hexagon circumvents the need for GPU-sized register files, allowing power conservation during lightweight tasks.

HVX also furnishes each thread with four 128-bit predicate registers, serving as storage for vector compares results and facilitating conditional accumulates through masking for select instructions.

While previous Hexagon iterations emphasized vector integer operations exclusively, Qualcomm has broadened HVX's scope by integrating floating-point capability, enhancing its versatility.

Given the typically sizable memory requirements of vectorized applications, Qualcomm forgoes caching vector accesses in L1, opting instead for L2 cache utilization as the primary cache level. Additionally, Hexagon boasts a sizable Tightly Coupled Memory (TCM), akin to AMD GCN's LDS but on a larger scale. In Snapdragon 8 Gen 2, Hexagon boasts an 8 MB TCM, pivotal for high-performance scatter and gather operations.

TCM proves crucial for operations like gather instructions, efficiently fetching data from non-contiguous memory locations, and scatters, which distribute data in the opposite direction. Since scatter and gather operations can be challenging for caches due to their memory access patterns, Hexagon avoids attempting them on cacheable memory, instead opting for TCM.

Similar to the scalar side, HVX houses specialized DSP hardware, featuring instructions like histogram calculations, capable of processing brightness values in an image, along with various other vector operations akin to AVX, including vector adds, min/max, absolute value, and reduction operations.

7.7.6 Tensor

HVX initially offered moderate performance for matrix multiplication in machine learning tasks. However, recognizing optimization potential, Qualcomm, like Nvidia, AMD, and Intel, sought to enhance its efficiency. Through the implementation of specialized matrix multiplication instructions, Hexagon DSP can now handle more computations per instruction, resulting in reduced power consumption.

To further improve performance, Qualcomm introduced a tensor coprocessor to Hexagon, aimed at accelerating machine learning tasks. This addition, despite its significant impact, prompted curiosity regarding its nomenclature—perhaps "Qualcomm Rectangle" would have been more fitting, given the matrix-like operations it performs.

Hexagon's NPU showcases remarkable capability, completing a substantial 16K multiply-accumulate operations per cycle, likely utilizing 4-bit weights. The Snapdragon 855/865 clocked their Hexagon DSP at 576 MHz. If the DSP can maintain the same clock speeds with the tensor unit activated, it should be capable of reaching 18.8 TOPS.

7.7.7   Performance

According to Wikipedia's disclosure (Wikipedia, 2024), Qualcomm's Snapdragon 855 contains their fourth generation on-device AI engine, which includes the Hexagon 690 DSP and Hexagon Tensor Accelerator (HTA) for AI acceleration. Snapdragon 865 contains the fifth generation on-device AI engine based on the Hexagon 698 DSP capable of 15 trillion operations per second (TOPS).

7.7.8   Brief

Hexagon offers a glimpse into the intricate domain of DSPs, positioning itself between CPU and GPU realms. It amalgamates expansive vector and tensor units with CPU features such as branch prediction, striking a balance between the two worlds. Its capabilities in thread-level parallelism also find a middle ground. Equipped with 6-way SMT, Hexagon surpasses typical CPU cores in thread-level parallelism, yet it lacks the extensive register files characteristic of GPU execution units, which handle multiple threads concurrently. Operating on a 4-wide VLIW execution model, Hexagon aims for a compromise. While recent GPUs tend toward scalar architectures with occasional dual issue capability, CPUs favor wide out-of-order execution. VLIW sacrifices some flexibility compared to out-of-order execution but allows for higher per-thread performance with well-structured code.

Consequently, the resultant processor offers CPU comforts while heavily relying on its instruction set to streamline hardware operations. This blend proves ideal for lightweight DSP tasks like audio decoding, where predictability favors VLIW performance, and the level of parallelism doesn't justify GPU execution. For a period, a "scalar" unit sufficed for Hexagon's needs.

Hexagon block diagram is shown in Fig. 7.33

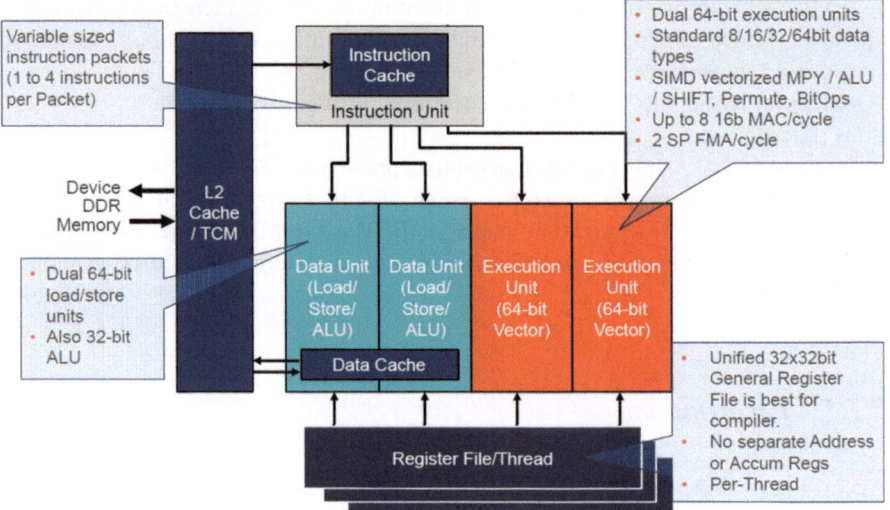

Fig. 7.33 Hexagon block diagram. From Qualcomm

With increasing processing demands, Qualcomm introduced HVX to enhance DSP throughput. Recognizing that not all DSP tasks require wide vector execution, the approach of allowing threads to acquire and release the HVX coprocessor made practical sense. Qualcomm highlighted image processing as a primary application for Hexagon, and the escalating resolutions of cell phone cameras likely validated the need for HVX.

Now, propelled by the rise of machine learning, Qualcomm has integrated a tensor unit into Hexagon. Similar to HVX, the tensor unit operates as a coprocessor that threads must request access to. While specifics on register file sizes were not provided by Qualcomm, their Hot Chips presentation primarily focused on the tradeoffs associated with various data types. Nevertheless, the inclusion of a tensor unit underscores Hexagon's adaptability.

7.8 Comparative Analysis of Embedded AI Accelerator Chips

Table 7.10 summarizes the above-mentioned main embedded AI accelerators from aspects such as AI inference performance, power consumption, and technical architecture.

Table 7.10 Comparison of mainstream embedded AI accelerators

Manufacturer	AI accelerator	Technology architecture	AI inference performance	Power (W)	Inference performance per W
NVIDIA	Jetson Nano	GPU	472 GFLOPS 1.88 TOPS	5 W/10 W (complete machine)	N/A
NVIDIA	Jetson TX2	GPU	1.33 TFLOPS	7.5 W/15 W (complete machine)	N/A
NVIDIA	Jetson X AVIER NX	GPU	6 TFLOPS 14/21 TOPS	10 W/15 W (complete machine)	N/A
NVIDIA	Jetson AGX XAVIER	GPU	32 TOPS	10 W/15 W/30 W (complete machine)	N/A
Intel	Myriad	VPU (V LIW architecture that mixes RISC, DSP and GPU)	1 TOPS	1 W	1 TOPS/W
Google	Edge TPU	TPU (ASIC)	4 TOPS	2 W	2 TOPS/W
XILINX	Z7020 + B1152x1	DPU (FPGA)	230 GOPS	2 W	0.115 TOPS/W
XILINX	ZU9 + B4096x3	DPU (FPGA)	4.1 TOPS	15 W	0.273 TOPS/W
ARM	Ethos-N77	NPU (ASIC)	4 TOPS	0.8 W	5 TOPS/W
Qualcomm	Hexagon 698	DSP	15 TOPS	6 W	2.5 TOPS/W

NVIDIA Jetson AGX XAVIER has the highest inference performance, while ARM Ethos-N77 has the highest performance/power ratio. Embedded system developers can choose appropriate AI accelerators based on inference performance and power requirements.

It is worth noting that the NVIDIA Jetson series is a complete AI computer. In addition to the GPU, it also includes the CPU, memory, and peripheral interfaces. Its power consumption is calculated based on the entire machine (GPU consumes the largest amount of power), while the power consumption of other AI accelerators is calculated based on the chip. In addition, the NVIDIA Jetson series has floating-point number inference capabilities, while other AI accelerators are based on integers for inference. Floating-point inference power is measured in GFLOPS, while integer inference power is measured in TOPS.

The following is the measured data of some of the AI accelerator inference performance. In this test, a Raspberry Pi board was used as a baseline, comparing it to a Raspberry Pi board with an Intel Movidius Compute Stick or Google Edge TPU's Oral USB Compute Stick installed, NVIDIA AI inference performance between Jetson Nano development board and Coreal development board with Google Edge TPU installed. Raspberry Pi is a common embedded development board, and its choice is representative. Versions 3.0 and 4.0 were used in testing. The USB interface of version 3.0 is version 2, while the USB interface of version 4.0 is version 3.

Inference performance is calculated in FPS, the processing speed (frames/s) of a video stream by a neural network model running on an AI accelerator (Table 7.11).

Among them, Jetson Nano is the only single-board computer that supports floating-point GPU acceleration. It supports most models since all frameworks (such as TensorFlow, Caffe, PyTorch, YOLO, MXNet, etc.) use the CUDA GPU support library. As with other AI accelerators, not all models can run, most often due to insufficient memory or incompatible hardware and/or software.

For different models, different accelerators show different inference capabilities. There is no accelerator that has the strongest inference capabilities for all models. This is caused by different chips having different instruction sets. For example, Intel Movidius has very strong inference performance on ResNet-50, while Google Edge TPU has very strong inference performance on MobileNet-V2. Overall, Jetson Nano supports the most models and has strong inference capabilities for them all.

Table 7.11 Comparison of measured performance of various embedded AI accelerators (source: https://qengineering.eu/deep-learning-with-raspberry-pi-and-alternatives.html#Compare_Jetson)

Model	Training framework	Raspberry Pi (TF-Lite)	Raspberry Pi + Intel Neural Stick 2	Raspberry Pi + Google Coral (Edge TPU) USB	Jetson Nano	Google Coral (Edge TPU)
EfficientNet-B0 (224 × 224)	TensorFlow	14.6 FPS (Pi3) 25.8 FPS (Pi4)	95 FPS (Pi3) 180 FPS (Pi4)	105 FPS (Pi3) 200 FPS (Pi4)	216 FPS	200 FPS
ResNet-50 (244 × 244)	TensorFlow	2.4 FPS (Pi3) 4.3 FPS (Pi4)	16 FPS (Pi3) 60 FPS (Pi4)	10 FPS (Pi3) 18.8 FPS (Pi4)	36 FPS	18.8 FPS
MobileNet-v2 (300 × 300)	TensorFlow	8.5 FPS (Pi3) 15.3 FPS (Pi4)	30 FPS (Pi3)	46 FPS (Pi3)	64 FPS	130 FPS
SSD Mobilenet-V2 (300–300)	TensorFlow	7.3 FPS (Pi3) 13 FPS (Pi4)	11 FPS (Pi3) 41 FPS (Pi4)	17 FPS (Pi3) 55 FPS (Pi4)	39 FPS	48 FPS
Binary model (300 × 300)	XNOR	6.8 FPS (Pi3) 12.5 FPS (Pi4)	–	–	–	–
Inception V4 (299 × 299)	PyTorch	–	–	3 FPS (Pi3)	11 FPS	9 FPS
Tiny YOLO V3 (416 × 416)	Darknet	0.5 FPS (Pi3) 1 FPS (Pi4)	–	–	25 FPS	–
OpenPose (25 × 256)	Caffe	4.3 FPS (Pi3) 10.3 FPS (Pi4)	5 FPS (Pi3)	–	14 FPS	–
Super Resolution (481 × 321)	PyTorch	–	0.6 FPS (Pi3)	–	15 FPS	–
VGG-19 (224 × 224)	MXNet	0.5 FPS (Pi3) 1 FPS (Pi4)	5 FPS	–	10 FPS	–
Unet (1 × 512 × 512)	Caffe	–	5 FPS	–	18 FPS	–
Unet (3 × 257 × 257)	TensorFlow	2.0 FPS (Pi3) 3.6 FPS (Pi4)				

References

ARM Limited. (n.d.). *AI platform for machine learning*. Retrieved from https://www.arm.com/products/silicon-ip-cpu/ai-platform

clamchowder. (2023). *Qualcomm's Hexagon DSP, and now, NPU*. Retrieved from https://chipsandcheese.com/2023/10/04/qualcomms-hexagon-dsp-and-now-npu/

Google LLC. (n.d.-a). *Edge TPU*. Retrieved from https://cloud.google.com/edge-tpu

Google LLC. (n.d.-b). *Products | Coral*. Retrieved from https://coral.ai/products/#prototyping-products

Intel Corporation. (n.d.). *Intel Movidius Vision Processing Units (VPUs)*. Retrieved from https://www.intel.com/content/www/us/en/products/details/processors/movidius-vpu.html

NVIDIA Corporation. (2018a). *NVDLA primer*. Retrieved from http://nvdla.org/primer.html

NVIDIA Corporation. (2018b). *NVIDIA Volta Architecture*. Retrieved from https://www.olcf.ornl.gov/wp-content/uploads/2018/12/summit_workshop_Volta-Architecture.pdf

NVIDIA Corporation. (2019). *Introducing Jetson Xavier NX, the World's Smallest AI Supercomputer*. Retrieved from https://developer.nvidia.com/blog/jetson-xavier-nx-the-worlds-smallest-ai-supercomputer/

NVIDIA Corporation. (n.d.). *Jetson modules*. Retrieved from https://developer.nvidia.com/embedded/jetson-modules

Q-engineering. (2021). *Google Coral Edge TPU explained in depth*. Retrieved from https://qengineering.eu/google-corals-tpu-explained.html

WikiChip. (2020). *Machine Learning Processor (MLP)—Microarchitectures—ARM*. Retrieved from https://en.wikichip.org/wiki/arm_holdings/microarchitectures/mlp

WikiChip. (2021). *SHAVE v2.0—Microarchitectures—Intel Movidius*. Retrieved from https://en.wikichip.org/wiki/movidius/microarchitectures/shave_v2.0

Wikipedia. (2024). *Qualcomm_Hexagon*. Retrieved from https://en.wikipedia.org/wiki/Qualcomm_Hexagon

XiLinx. (2017). *Deep learning with INT8 optimization on Xilinx devices*. Retrieved from https://docs.amd.com/v/u/en-US/wp486-deep-learning-int8

XiLinx. (n.d.) *AI inference acceleration*. Retrieved from https://www.xilinx.com/applications/megatrends/machine-learning.html

Chapter 8
Software Framework for Embedded Neural Networks

Abstract This chapter introduces common embedded neural network software frameworks in detail. Some of them are universal and can be adapted to various AI acceleration chips, such as TensorFlow Lite, Apache TVM, etc., while others are developed for specific AI acceleration chips, such as TensorRT, OpenVINO, XILINX Vitis, uTensor, Qualcomm AI stack, etc. For each framework, its main functions, module composition, workflow, etc. are introduced. Finally, each framework is compared in terms of supported AI chip types and supported neural network training frameworks to facilitate developers to choose according to their needs.

As AI becomes a major driver of edge technologies, the combination of hardware accelerators and software platforms becomes increasingly important for running inference models. Some hardware accelerators have launched supporting software development frameworks, such as NVIDIA's TensorRT. There are also some manufacturers that have launched general embedded neural network software frameworks, such as TensorFlow Lite. They are introduced separately below.

Keywords TensorFlow Lite · TensorRT · OpenVINO · XILINX Vitis · uTensor · Apache TVM · Qualcomm AI Stack

8.1　TensorFlow Lite

TensorFlow Lite (TensorFlow, n.d.) is an open-source deep learning framework for device-side inference launched by Google based on TensorFlow.

TensorFlow Lite is a set of tools that help developers run TensorFlow models on mobile, embedded, and IoT devices. It supports on-device machine learning inference, has low latency, and has small binaries.

8.1.1　Introduction to TensorFlow Lite

TensorFlow Lite consists of two main components:

© Tsinghua University Press 2024

B. Li, *Embedded Artificial Intelligence*,

https://doi.org/10.1007/978-981-97-5038-2_8

- **TensorFlow Lite interpreter**, which can run specially optimized models on many different types of hardware, including mobile phones, embedded Linux devices, and microcontrollers.
- **TensorFlow Lite translator**, which converts TensorFlow models into an efficient form for use by the interpreter and introduces optimizations to reduce binary size and improve performance.

TensorFlow Lite is designed to make it easy for you to perform machine learning on devices at the "edge" of your network, without having to send data back and forth between the device and the server. For developers, performing machine learning on-device can help:

- Reduced latency: Data does not need to travel to and from the server.
- Protect privacy: No data leaves the device.
- Reduced connection: No internet connection required.
- Reduce power consumption: Network connections are very power-hungry.

TensorFlow Lite supports devices ranging from ultra-small microcontrollers to powerful mobile phones. For example, the TensorFlow Lite binary is approximately 1 MB in size (for a 32-bit ARM build) when using all 125 supported operators; when using only the operators required to support common image classification models (InceptionV3 and MobileNet), the TensorFlow Lite binary is less than 300 KB in size.

The main features of TensorFlow Lite are

- An interpreter tuned for on-device machine learning, supporting a core set of operators optimized for on-device applications and with smaller binaries.
- Supports multiple platforms (covering Android and iOS devices, embedded Linux, and microcontrollers) and leverages platform APIs to perform accelerated inference.
- Supports APIs in multiple languages, including Java, Swift, Objective-C, C++, and Python.
- High performance performs hardware acceleration on supported devices and provides device-optimized kernels and pre-fused activation functions and biases.
- Model optimization tools (including quantization) to reduce model size and improve performance without impacting accuracy.
- Efficient model format using FlatBuffer optimized for small device size and portability.
- Pre-trained models for common machine learning tasks that can be customized for your application.

The workflow using TensorFlow Lite includes the following steps, as shown in Fig. 8.1.

1. Select a model

   You can use your own TensorFlow model, or find a model online, or choose a model from our pre-trained models to use directly or retrain.

Pick a model	Convert	Deploy	Optimize
Pick a new model or retrain an existing one.	Convert a TensorFlow model into a compressed flat buffer with the TensorFlow Lite Converter.	Take the compressed .tflite file and load it into a mobile or embedded device.	Quantize by converting 32-bit floats to more efficient 8-bit integers or run on GPU.

Fig. 8.1 TensorFlow Lite workflow

2. Conversion model

   If you are using a custom model, use the TensorFlow Lite converter to convert the model to TensorFlow Lite format with just a few lines of Python code.
3. Deploy to your device

   Run your model on-device using the TensorFlow Lite interpreter, which has an API that supports multiple languages.
4. Optimize your model

   Use our model optimization toolkit to reduce model size and increase efficiency while minimizing the impact on accuracy.

TensorFlow Lite plans to provide high-performance device-side inference for any TensorFlow model. However, the TensorFlow Lite interpreter currently only supports a limited number of TensorFlow operators, which are optimized for on-device use cases. That said, some models require additional steps to use TensorFlow Lite.

If the TensorFlow Lite interpreter does not yet support an operator used by your model, you can add TensorFlow operations using TensorFlow Select in a TensorFlow Lite build. However, this increases the size of the binary file.

TensorFlow Lite currently does not support on-device training.

8.1.2   How TensorFlow Lite Works

TensorFlow Lite takes the generated model (frozen graph, SavedModel or HDF5 model) as input, packages and deploys it, and then interprets it in the client application, achieving resource saving and optimization in the process, as shown in Fig. 8.2.

TensorFlow Lite consists of the following components, as shown in Fig. 8.3.

Below, we take a look at some key optimizations of TensorFlow Lite components.

1. Model converter

   The TensorFlow Lite Converter (TOCO) takes a trained TensorFlow model as input and outputs a TFLite (.tflite) file, which is a FlatBuffer-based file that contains a simplified binary representation of the original model.

   FlatBuffers can efficiently serialize model data and provide fast access to that data while maintaining a small binary size. This is particularly useful for model files that are populated with large amounts of numeric weight data, which may cause significant latency in read operations due to their size.

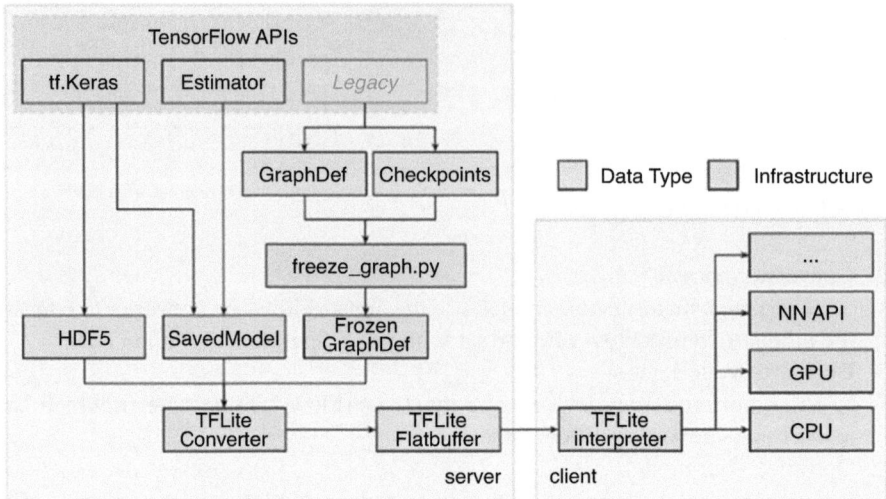

Fig. 8.2 TensorFlow Lite internal working process

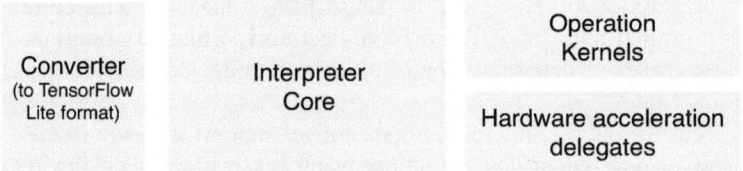

Fig. 8.3 Components of TensorFlow Lite

Using the FlatBuffer protocol as the basis for model conversion, TensorFlow Lite avoids inefficient file parsing and unparsing operations that can lead to slower execution.

2. Interpreter core

TensorFlow Lite models in client applications using a simplified set of TensorFlow operators. By limiting the default operators, libraries, and tools required to run Lite models, the core of the interpreter is reduced to about 100 kb, and all supported kernels are only about 300 kb.

If a model requires operators outside of the provided set, TensorFlow Lite allows the implementation of custom operators. This optional approach is key to keeping TensorFlow lean.

3. Hardware acceleration

TensorFlow Lite's optimizations extend all the way to the hardware. Working within the tight constraints of mobile and embedded devices means processors must be utilized to ultra-efficient standards.

The Android NDK includes a set of Neural Network APIs (NNAPI) that provide access to hardware-accelerated inference operations on Android devices.

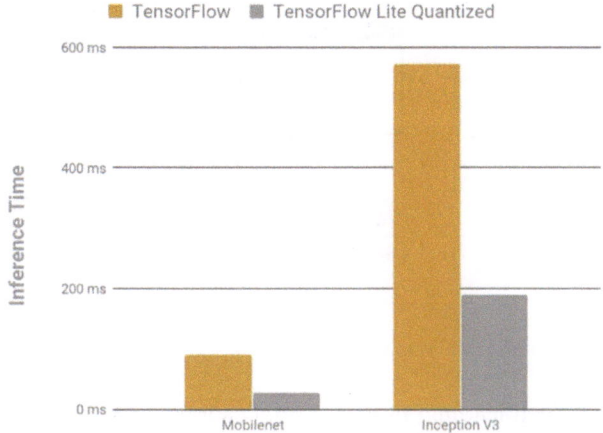

Fig. 8.4 Comparison of inference speed before and after quantization in TensorFlow Lite

NNAPI works with TensorFlow Lite to use hardware accelerators where available to find the best model operations. As machine learning hardware becomes more prevalent on edge devices, the benefits of the NNAPI framework will become even more apparent.

In addition, TensorFlow Lite allows a subset of models and operations to selectively use the GPU on embedded devices. Improve speed and efficiency for models that work too much in parallel and suffer from loss of quantization accuracy. For some neural networks, the efficiency improvement is up to seven times.

4. Quantization

Quantization refers to the process of reducing a set of consecutive numbers to a set of smaller consecutive numbers without losing the descriptiveness of the initial set. In the context of neural networks, this usually means reducing the precision of operations from 32-bit floating-point numbers to 8-bit values.

TensorFlow Lite encourages post-training quantization and is provided as an attribute of the TOCO converter during the conversion step.

Benchmark tests show that compressed model inference latency can be reduced by up to three times with negligible degradation in inference accuracy, as shown in Fig. 8.4.

Taken together, these core optimizations provide a reliable framework that provides strong support for implementing artificial intelligence on embedded devices.

8.2 TensorRT

NVIDIA TensorRT (NVIDIA Corporation, 2018) is an SDK for high-performance deep learning inference. It includes a deep learning inference optimizer and runtime environment, and its core is a C++ library that can provide low latency and

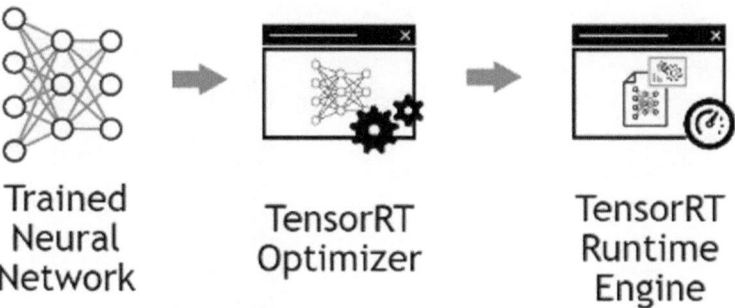

Fig. 8.5 TensorRT optimizer and runtime engine

high-throughput deep learning inference applications on NVIDIA GPU, as shown in Fig. 8.5.

During inference, TensorRT-based applications run up to 40 times faster than CPU-only platforms. With TensorRT, neural network models trained in various mainstream frameworks can be optimized, accurately calibrated to low precision, and finally deployed to embedded hardware platforms.

TensorRT is built on NVIDIA's parallel programming model CUDA, leveraging CUDA-X libraries for artificial intelligence, automata, high-performance computing, and graphics to optimize inference for all deep learning frameworks.

TensorRT provides 8-bit integer and 16-bit floating-point optimization for production deployment of deep learning inference applications, such as video recognition, speech recognition, recommendation systems, and natural language processing. Reduced-precision inference significantly reduces application latency, which is required for many real-time services, autonomous vehicles, and embedded applications.

It can work in a complementary manner with training frameworks such as TensorFlow, Caffe, PyTorch, MXNet, etc. It specializes in running trained networks on GPUs quickly and efficiently to generate results (a process called scoring, detection, regression, or inference depending on the scenario).

Some training frameworks, such as TensorFlow, have integrated TensorRT, so it can be used to accelerate inference within the framework. Additionally, TensorRT can be used as a library in user applications. It includes parsers for importing existing models from Caffe, ONNX or TensorFlow, as well as C++ and Python APIs for building models programmatically.

Trained models can be imported into TensorRT from various deep learning frameworks. After optimization, TensorRT selects specific cores based on the platform, thereby improving inference performance in data centers, embedded platforms, and autonomous driving platforms.

Specifically, TensorRT achieves performance optimization in the following ways, as shown in Fig. 8.6.

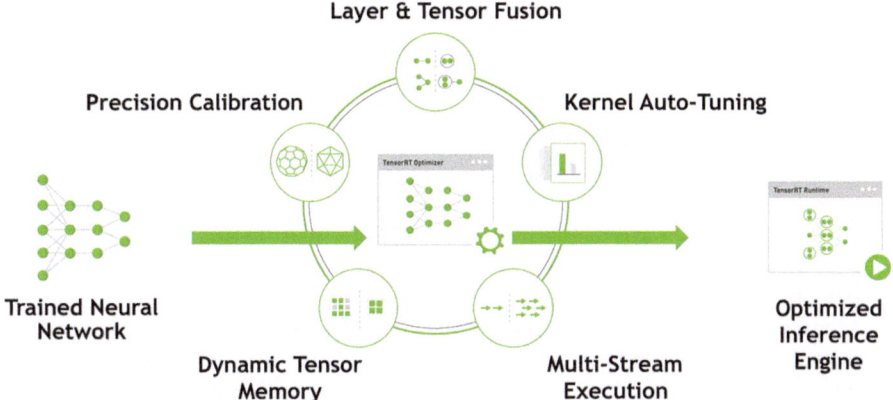

Fig. 8.6 How TensorRT works

1. Accuracy calibration of weights and activation values: Improve throughput while maintaining accuracy by quantizing the model into 8-bit integers.
2. Layer and tensor fusion: Optimize GPU memory and bandwidth usage by fusing nodes in the kernel.
3. Kernel automatic adjustment: Select the best data layer and algorithm according to the target GPU platform.
4. Dynamic tensor memory: Minimize memory usage and effectively reuse tensor memory.
5. Multiple input streams in parallel: With scalable design, multiple input streams can be processed in parallel.

   In short, TensorRT greatly improves the performance of deep learning inference on NVIDIA GPUs, meeting the inference needs of embedded environments with high throughput and ultra-low latency.

8.2.1 Benefits of TensorRT

After the neural network is trained, TensorRT can be used as a runtime breakaway framework to compress, optimize, and deploy the network.

TensorRT optimizes matrix math operations by merging layers according to specified precision (FP32, FP16, or INT8), optimizing kernel selection, and performing normalization and transformations to improve latency, throughput, and efficiency.

For deep learning inference, there are five key factors to measure:

1. Throughput: The amount of output within a given time range. Throughput per server, typically measured in inferences/second or samples/second, is critical for cost-effective scaling of data centers.

2. Efficiency: Throughput delivered per unit of power, usually expressed as performance/Watt. Efficiency is another key factor in cost-effectively scaling a data center, as servers, racks, and entire data centers must operate within a certain power budget.
3. Latency: The time to perform inference, usually in milliseconds. Low latency is critical to delivering the rapidly growing number of real-time inference-based services.
4. Accuracy: The ability of a trained neural network to provide correct answers. For image classification-based usage, key metrics are expressed as TOP-5 or TOP-1 percentage.
5. Memory occupation: The memory used for network inference on the host and device, its size depends on the algorithm used. This limits which networks, and which combinations of networks, can run on a given inference platform. This is especially important for systems that require multiple networks and have limited memory resources, such as cascaded multi-class detection networks used in intelligent video analysis, and multi-camera, multi-network autonomous driving systems.

Alternatives to TensorRT include

- Perform inference using the training framework itself.
- Write custom applications specifically designed to perform networking using low-level libraries and mathematical operations.

It's easy to perform inference using a training framework, but the performance on a given GPU is often much lower compared to using an optimized solution like TensorRT. The training framework implements more general code that emphasizes versatility and is optimized for efficient training.

Writing custom applications specifically for executing neural networks can lead to greater efficiency, however, this can be very laborious and requires considerable expertise to achieve high-performance levels on the GPU. Additionally, optimizations that run on one GPU may not fully translate to other GPUs, and each generation of GPUs may introduce new features that can only be exploited by writing new code.

TensorRT solves the above problems by highly abstracting specific hardware details into an API and specifically developing and optimizing it for high throughput, low latency, and low memory footprint inference.

8.2.2   TensorRT Main Functions

TensorRT enables developers to import, calibrate, generate, and deploy optimized networks. Networks can be imported directly from Caffe or from other frameworks via UFF or ONNX formats. It can also be created programmatically, i.e., instantiating individual layers, and setting parameters and weights directly.

Users can also run custom layers through TensorRT using the plug-in interface. The GraphSurgeon utility provides the ability to map TensorFlow nodes to custom layers in TensorRT, allowing inference on TensorFlow networks using TensorRT.

TensorRT provides C++ implementations on all supported platforms, and Python implementations on x86, aarch64, and ppc64le.

The key interfaces in the TensorRT core library are

1. Network definition: The network definition interface provides applications with a way to define a network. It can specify input and output tensors, can add layers, and has an interface for configuring various types of layers. Layer types such as convolutional layers and recurrent layers, as well as plug-in layer types, allow applications to implement functionality that TensorRT itself does not support.
2. Builder: The Builder interface allows the creation of optimized engines based on network definitions. It allows applications to specify maximum batch and work-space sizes, minimum acceptable precision levels, timed iteration counts for automatic tuning, and an interface for quantizing the network to run with 8-bit precision.
3. Engine: The Engine interface allows applications to perform inference. It supports synchronous and asynchronous execution, performance analysis, and input and output of enumeration and query engines. A single engine can have multiple execution contexts, thereby executing multiple batches of a set of trained parameters.

TensorRT provides a parser for importing trained networks to create network definitions:

1. Caffe parser: This parser is used to parse Caffe networks created in BVLC Caffe or NVCaffe 0.16. It also provides the ability to register plug-ins for custom layers.
2. UFF parser: This parser is used to parse networks in UFF format. It also provides the ability to register plug-ins for custom layers and pass field properties.
3. ONNX parser: This parser is used to parse ONNX models. ONNX is a standard for representing deep learning models and an open model format. PyTorch, MXNet, TensorFlow, Caffe2, Chainer, CNTK, PaddlePaddle, and many other frameworks support the ONNX format. It can thus be used to convert model formats between various training frameworks.

8.2.3   How Tensor RT Works

To optimize an inference model, TensorRT takes a network definition, performs (platform-specific) optimizations, and generates an inference engine. This process is called the build phase. The build phase can take a lot of time, especially when running on an embedded platform. Therefore, a typical application builds the engine once and then serializes it into a scenario file for later use.

Note The generated solution files are not portable across platforms or TensorRT versions. Scheme files are specific to the exact GPU model they are built on and must be rebuilt if you want to run on a different GPU.

The build phase performs the following optimizations on the graph:

- Eliminate unused layers from the output.
- Eliminate actions that are equivalent to no actions.
- Fusion of convolution, bias and ReLU operations.
- Aggregate operations with sufficiently similar parameters and identical source tensors (e.g., 1x1 convolutions in the GoogleNet v5 inception module).
- Merge concatenated layers to direct layer output to the correct destination.

The builder can also modify the precision of the weights if necessary. When a network is generated with 8-bit integer precision, it uses a process called calibration to determine the dynamic range of intermediate activation values and thus the appropriate quantization scaling factor.

Additionally, the build phase runs dummy data over the layers to select the fastest layer and preformat weights or optimize memory where appropriate.

In Sect. 11.2, the working method of TensorRT will be introduced in depth and will not be described in detail here.

These are the basics of how to use TensorRT to optimize deep learning applications for inference. There are many other optimization techniques, such as overlapping data transfers between the CPU and GPU while calculating, and using 8-bit integer precision, that can help you achieve higher inference performance. Overall, by using TensorRT to move inference from the CPU to the GPU, latency can be reduced by a factor of 100.

8.3   OpenVINO

8.3.1   Introduction to OpenVINO

The demand for smart vision solutions increases. This requires the development of tools that integrate computer vision, deep learning, and analytical processing capabilities into applications to transform data into insights and drive artificial intelligence. Intel Distribution of OpenVINO toolkit, referred to as OpenVINO, is a free development kit launched by Intel. It is a toolkit for quickly developing high-performance computer vision and integrating deep learning into vision applications. It ensures that deep learning operates on hardware accelerators, OpenVINO™ Through heterogeneous computing, the powerful performance of Intel hardware platforms (including CPU, GPU, Intel FPGA, and Intel Movidius VPU) can be fully utilized, and layered execution can be performed across Intel platforms. By using the OpenVINO toolkit, developers can implement deep learning inference in application scenarios such as image surveillance, retail, production, smart cities, healthcare, office automation, and transportation, bringing substantial performance improvements to deep learning workloads.

1. Software tools to accelerate deep learning inference

   The Intel OpenVINO toolbox is Movidius' default software development kit for optimizing performance, integrating deep learning inference, and running deep neural networks (DNN) on the Intel® Movidius™ Vision Processing Unit (VPU). The toolkit supports a wide range of neural networks, simplifying deployment not only between NCS 2 hardware but across Intel Vision Accelerator solutions.

2. Develop on one platform and deploy across multiple platforms

   This is the slogan and brief description of the Intel distribution of the OpenVINO toolbox. Thanks to the Intermediate Representation (IR) format, you can develop and test neural networks on one type of processor (e.g., CPU) and deploy the same on a range of processing units (e.g., CPU, GPU, V PU, FPGA) model, or even heterogeneous deployment across two processors (split model). IR concepts allow you to run models built using multiple frameworks, such as TensorFlow, Pytorch, and MXNet, as well as other exchange formats such as ONNX. Due to the modular architecture of the toolkit, it can flexibly support multiple frameworks, exchange formats and hardware accelerators. Figure 8.7 shows a simplified graphical representation of the OpenVINO toolbox software component.

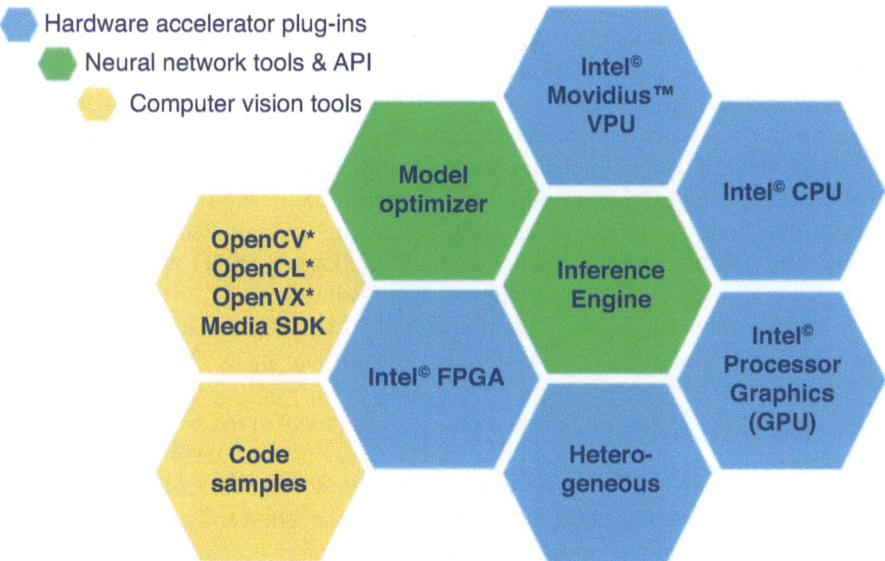

Fig. 8.7 Intel OpenVINO Toolkit

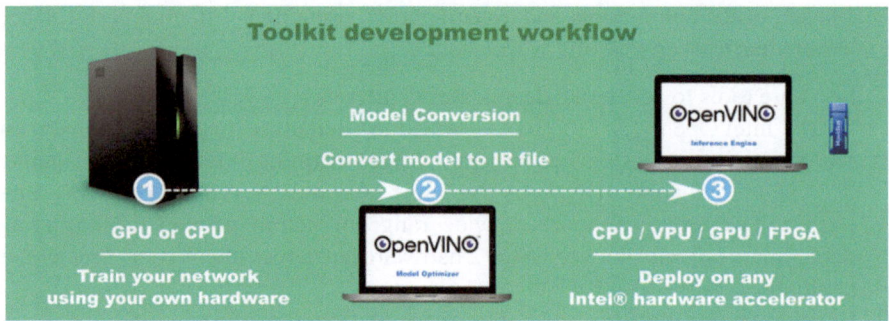

Fig. 8.8 OpenVINO toolkit workflow

3. Simplified development workflow

The toolkit features a simple development workflow that requires only three steps to develop and deploy neural networks on any supported processor and accelerator, as shown in Fig. 8.8.

(a) Use one of the supported frameworks to train the model on your preferred training hardware.
(b) Convert the trained model to an IR file using the toolbox's model optimizer.
(c) Load IR models to supported hardware accelerators to perform inference.

8.3.2   Structure of OpenVINO

The OpenVINO development kit contains two major components: Intel Deep Learning Deployment Toolkit (DLDT) and Intel's traditional computer vision toolkit combination, such as OpenCV, OpenVX, etc. OpenVINO uses DLDT to deploy deep learning inference models and uses the Intel Multimedia Software Development Kit (Intel Media SDK) in Intel's traditional computer vision toolkit to accelerate the encoding/decoding of videos and images, enhance the performance of image processors, and improve image processing. Processing is optimized and accelerated. Combining Intel's traditional computer vision toolkit and DLDT provides the ability to build high-performance, intelligent vision solutions.

Figure 8.9 shows the architecture diagram of DLDT. The two core suites used by the OpenVINO development kit for deep learning are in DLDT, namely: model optimizer (MO) and inference engine (IE).

The model optimizer (MO) imports the trained model in the preparation stage. It currently supports models based on a variety of mainstream deep learning frameworks, such as models based on Caffe, TensorFlow, MxNet, ONNX, and Kaldi. Then the input trained machine learning model is converted and optimized, and the model is converted into a hardware-matched data type to accelerate the subsequent inference performance. The output of the MO after model conversion is a set of files in the Intermediate Representation (IR) format. Files in IR format can be used to

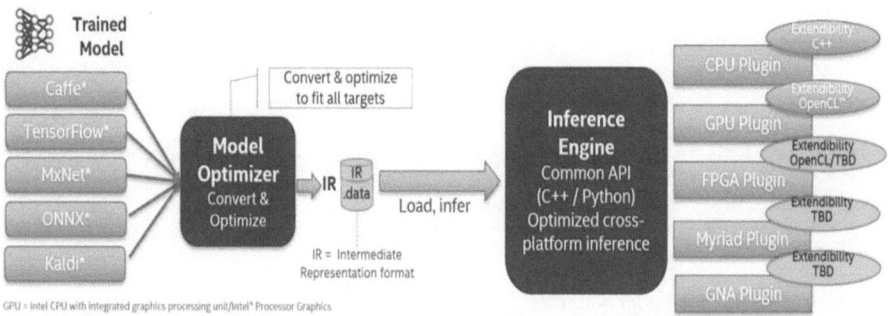

Fig. 8.9 Intel Deep Learning Deployment Toolkit Architecture Diagram

describe the entire model, which contains two types of files: .xml and .bin files. .xml stores files describing the network topology of the model and .bin is a binary file that saves important parameters such as model weights and biases.

Next, the IR format file will be sent to the inference engine (IE), and then calls Intel-related hardware to implement inference. IE has a unified API for developers to call. Currently supported hardware includes CPU, GPU, Intel FPGA, and Intel Movidius VPU. The OpenVX and OpenCV libraries in IE can also make traditional computer graphics-related development work such as image/video processing, computer vision tracking, and feature extraction faster and more efficient.

8.3.3   OpenVINO Application Development

There are four stages in the development cycle of AI applications. The entire development process is as follows: data collection and cleaning—creating a machine learning model—inputting data into the model for training and parameter tuning—after the trained model for inference and deployment. OpenVINO is designed for the fourth phase, and DLDT provides solutions for deployment from edge to cloud.

In the traditional deep learning application development and deployment process, the trained model, and data files such as videos/pictures/voices that need to be inferred are input into the user inference application to implement inference. After the introduction of OpenVINO, it is necessary to add a step of model optimization after the trained model and use MO to optimize and convert the model. This operation is a one-time operation, which means that only if the trained model changes, you need to execute MO again to obtain a new IR file. Finally, just like the traditional deployment process, the IR file, the trained model, and the video/picture/voice data files that need to be inferred are input into the user inference application to implement inference. In user inference applications, IE-related APIs need to be called to process these inferences.

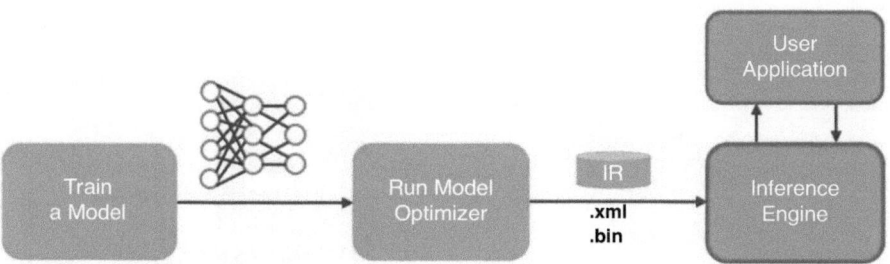

Fig. 8.10 OpenVINO model optimizer

Model Optimizer (MO)

MO currently supports a variety of mainstream deep learning frameworks. Through MO, models under these different frameworks can be optimized and converted into a unified IR form output model, as shown in Fig. 8.10.

MO performs many optimized operations in model conversion: such as node merging, batch data regularization, horizontal fusion, constant folding, etc.

Developers do not need to perform any development operations on MO. It is a command line tool. You only need to execute a command line statement to complete all model conversion work.

Inference Engine (IE)

IE provides a unified set of APIs across Intel hardware. After developers call these APIs, IE will automatically allocate corresponding plug-ins to implement inference. For example, when we want the user application to be executed on the CPU, IE will call the Intel MKL-DNN plug-in, and if we want to execute it on the GPU, it will call clDNN plug-in to calculate. These plug-ins will achieve better computing results after performance optimization on Intel architecture, as shown in Fig. 8.11.

IE supports heterogeneous computing and asynchronous computing of hardware. Heterogeneous computing refers to choosing different hardware to run on different layers. For example, if FPGA and CPU are selected as the parameters for heterogeneous computing, if when performing operations on a certain layer, it is found that the computing performance on the FPGA is not as good as on the CPU, then IE will feed back to the CPU for this layer. Operation. Through this mechanism, developers' user applications are guaranteed to have better performance. Asynchronous computing means that after making an inference request, there is no need to wait until the inference request is processed before executing the next work. Instead, while waiting for the execution result of the request, the next batch of data is put into the model for processing and makes an inference request for this batch of data. Execute sequentially executed tasks in parallel to improve performance.

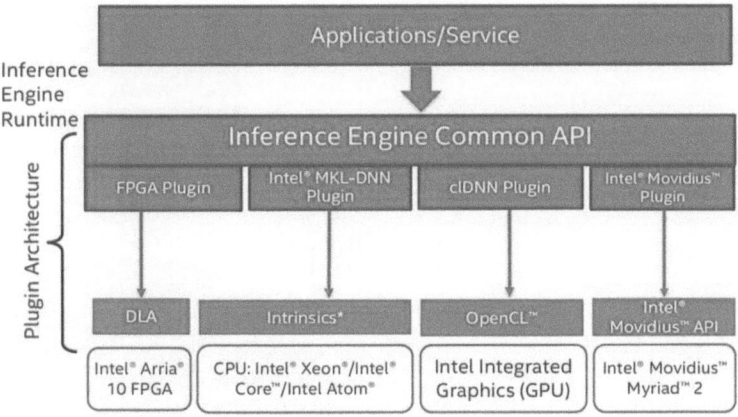

Fig. 8.11   OpenVINO inference engine

Workflow Based on Inference Engine Development

The workflow for developing applications based on the inference engine includes two parts, as shown in Fig. 8.12.

The first part is the initialization work. Initialization can be broken down into the following steps:

1. Load the model and corresponding parameters.
2. Set the batch size (if not set, the default size is 1).
3. Load inference plug-ins (such as CPU, GPU, FPGA, and VPU).
4. Load the model's network topology into the plug-in. Allocate input and output buffers.

The second part is the data filling operation. Input the data that needs to be inferred into the inference model. This is a cyclic operation:

1. Fill the input buffer with inference data.
2. Call IE's API to implement inference.
3. Output the inference results.

Following the above workflow, complete AI applications can be quickly developed and deployed to embedded devices.

8.4   XILINX Vitis

The Vitis AI development environment (XiLinx, n.d.) is XILINX's development platform for AI inference on XILINX hardware platforms (including edge devices and Alveo cards), as shown in Fig. 8.13. It consists of optimized IP, tools, libraries, models, and example designs. It is designed with high efficiency and ease of use in

Fig. 8.12 OpenVINO
inference engine workflow

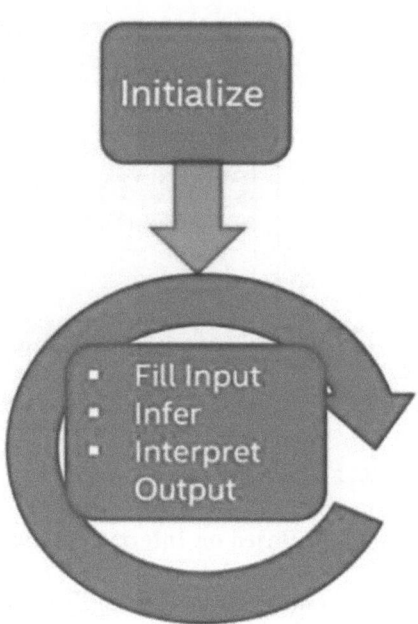

mind, fully utilizing the potential of AI acceleration on XILINX FPGA and ACAP. At the edge, Vitis AI acceleration delivers superior AI inference performance. It improves frame rate (FPS), reduces power consumption, and enables 5× to 50× network performance optimization.

XILINX Vitis has the following features:

- Supports mainstream frameworks and the latest models and can complete various deep learning tasks.
- Provides a comprehensive set of pre-optimized models ready for deployment on XILINX devices. The closest model can be found and started retraining for the application.
- Provides a powerful open-source quantizer that supports pruned and unpruned model quantization, calibration, and fine-tuning.
- AI Analyzer provides layer-by-layer analysis to help resolve bottlenecks.
- The AI library provides open-source high-level C++ and Python APIs for maximum portability from edge to cloud.
- Efficient and scalable IP cores can be customized to meet different needs from multiple perspectives such as throughput, latency, power consumption, and accuracy.

1. AI optimizer

    With model compression technology, Vitis can reduce model complexity by 5–50 times while minimizing the impact on accuracy. Currently, the Vitis AI optimizer only contains a tool called pruner. The Vitis Artificial Intelligence Pruner (VAI pruner) prunes redundant connections in neural networks, reducing

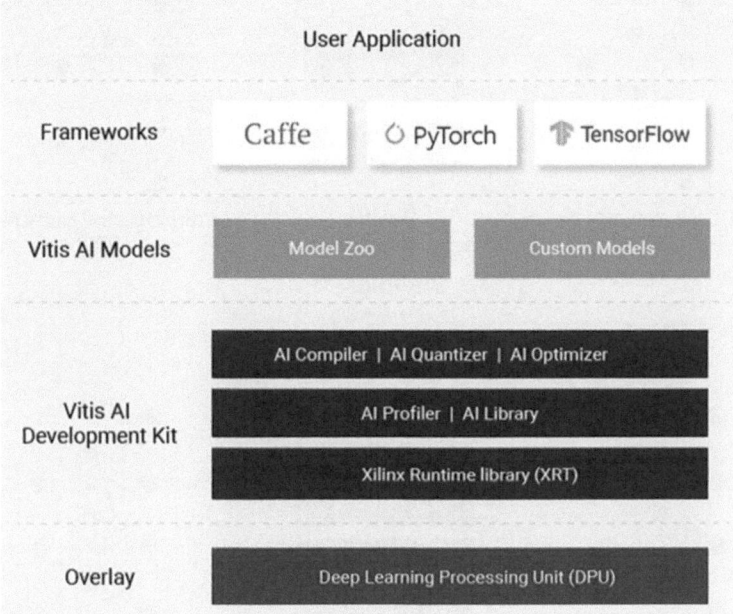

Fig. 8.13   XILINX Vitis AI Development Environment

the overall required operations. The pruned model produced by the AI pruner can be further quantized by the AI quantizer and deployed to the FPGA, as shown in Fig. 8.14.

2. AI quantizer

By converting 32-bit floating-point weights and activation values into fixed-point numbers (such as INT8), the AI quantizer can reduce computational complexity without losing prediction accuracy, as shown in Fig. 8.15. Fixed-point network models require less memory bandwidth and therefore provide faster speeds and greater power efficiency than floating-point models.

3. AI compiler

Map AI models to efficient instruction sets and data flows. Simultaneously perform complex optimizations such as layer fusion, instruction scheduling, and reuse on-chip memory as much as possible, as shown in Fig. 8.16.

4. AI analyzer

Performance analyzer provides in-depth analysis of the efficiency and utilization of AI inference implementations. Vitis AI analyzer analyzes and visualizes AI applications to find bottlenecks and allocate computing resources across different devices. It can track function calls and running time without changing the code, and can also collect hardware information, including CPU, DPU and memory utilization, as shown in Fig. 8.17.

5. AI library

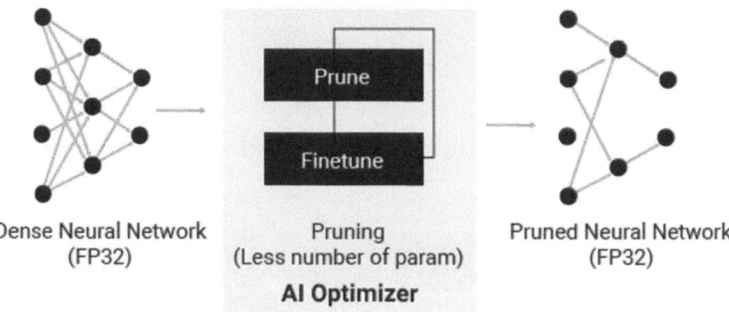

Fig. 8.14 XILINX AI optimizer

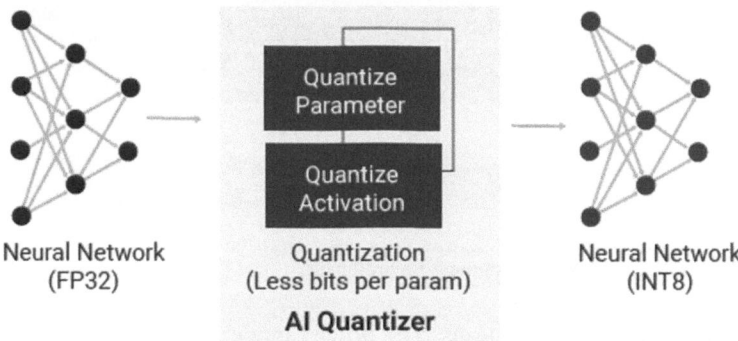

Fig. 8.15 XILINX Vitis AI Quantizer

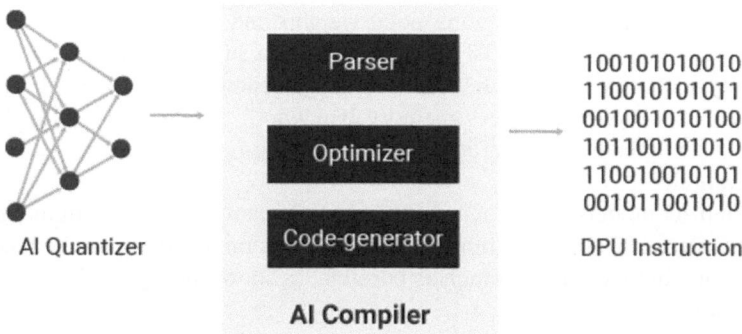

Fig. 8.16 XILINX Vitis AI Compiler

The Vitis AI library is a set of high-level libraries and APIs for efficient AI inference with deep learning processor units (DPUs). It is built on Vitis AI Runtime, has a unified API, and provides an easy-to-use interface for AI model deployment on the XILINX platform, as shown in Fig. 8.18.

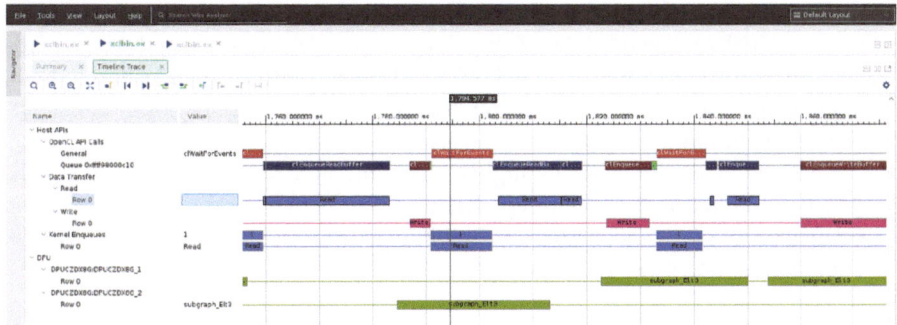

Fig. 8.17 XILINX Vitis AI analyzer

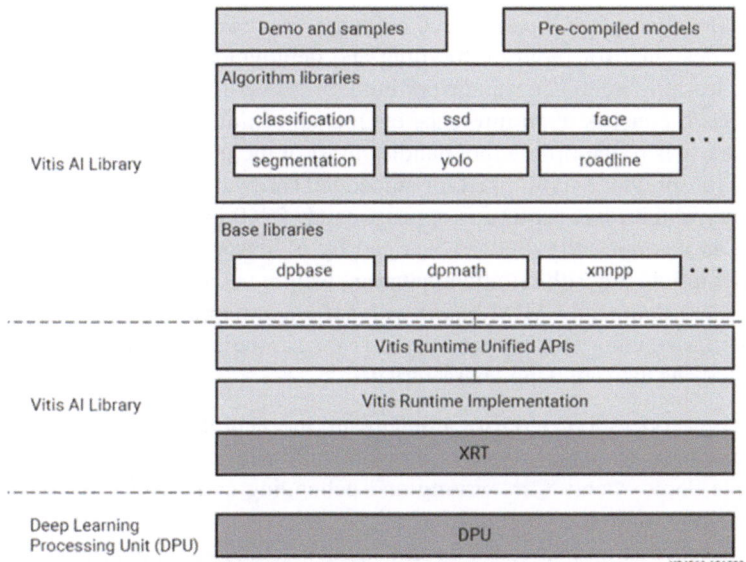

Fig. 8.18 XILINX Vitis AI Library

The Vitis AI library encapsulates many efficient, high-quality neural networks and provides an easy-to-use and unified interface. This simplifies the use of deep learning neural networks, even for users without deep learning or FPGA knowledge. The Vitis AI library allows users to focus more on developing applications rather than the underlying hardware.

6. DPU support

Vitis AI provides support for a range of different DPUs, embedded devices such as XILINX Zynq®-7000, Zynq® UltraScale +™ MPSoC , and Alveo cards such as U50, U200, U250, and U280, delivering performance in throughput, latency, scalability and unique differentiation and flexibility in power consumption are achieved.

In short, Vitis AI provides highly adaptable and real-time artificial intelligence inference acceleration for XILINX series AI chips. AI inference on embedded devices and the cloud is optimized through pruning, quantization, compilation, and other technologies.

8.5 uTensor

uTensor (n.d.) is a free, open-source software framework for embedded machine learning, designed for rapid prototyping and deployment. It is a very lightweight machine learning inference framework built on Tensorflow and optimized for ARM processors. It consists of an inference engine, an offline tool, and a data collection framework, where the offline tool handles most of the model transformation work.

The uTensor code is optimized for embedded hardware with only a few kilobytes of memory, and its core runtime is approximately 2 KB in size! Taking a three-layer MLP as an example, the model generated by u Tensor is only 32 KB, including model definition, algorithm implementation, weight values, etc.

uTensor supports any ARM board with sufficient memory (128+ KB RAM and 512 KB+ flash memory recommended) and Mbed as the embedded operating system.

uTensor has the following characteristics:

1. Real-time operation: uTensor implements the neural network model in C++, which can be manually coded or automatically generated from a trained model.
2. Simple deployment: Start writing and debugging code from Jupyter notebook and deploy to MCU.
3. Model processing: Innovation in embedded machine learning also requires rapid iteration. uTensor SDK provides easily customizable model converters via Python.
4. Open ecosystem: uTensor is an embedded machine learning framework. You can define your own operators, model converters, or port them to new platforms.

The workflow of uTensor is shown in Fig. 8.19.

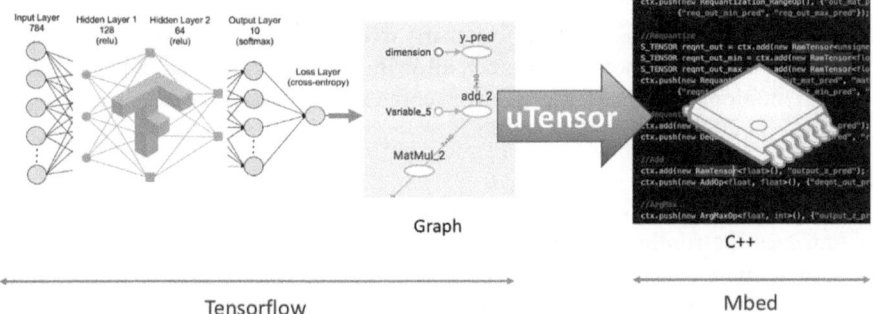

Fig. 8.19 uTensor workflow

First, build and train a model in Tensorflow. Next, uTensor uses the model and produces a .cpp and .hpp files that contain the generated C++11 code required for inference. Finally, using it on embedded devices.

```
#include "models/deep_mlp.hpp"
...
Context ctx; //creating a context
...
//preparing for the input tensor
...
get_deep_mlp_ctx (Context& ctx, Tensor* input_0); //perform
inference
ctx.eval ();
S_TENSOR prediction = ctx.get ({"y_pred:0"}); //getting
the result
```

The .hpp and .cpp files of uTensor model are generated from the model file, for example:

```
$ utensor -cli deep_mlp. pb --output-nodes= y_pred
...
...Generate weight file: models/deep_mlp_weight.hpp
...Generate header file: models/deep_mlp.hpp
...Generate source file: models/deep_mlp.cpp
```

The uTensor runtime consists of two main components: 1) the uTensor Core, which includes the essential data structures, interfaces, and types required for the uTensor performance runtime, and 2) the uTensor Library, a set of reasonable default implementations built on top of the uTensor Core. The build system compiles these two components separately, allowing users to easily extend and override implementations on top of the uTensor core, such as custom memory managers, tensors, operators, error handlers, etc.

1. uTensor core
   As its name suggests, the uTensor core is the central component that defines and enforces the runtime's strict guarantees, such as memory safety, model updatability, and a consistent user experience. Although it looks like a high-level language, uTensor core takes up very little space after compilation, about 1–2 KB (in addition, the uTensor library size is about 1 KB).
2. uTensor library
   The uTensor library is a set of sensible default implementations built on top of the uTensor core. These include

   (a)   Error handler.
   (b)   Configurator.
   (c)   Context.
   (d)   Operator.
         Legacy operator.
         Optimized operator.
         Symmetric quantization operator.
         Reference operator.
   (e)   Tensor.

   Currently, uTensor supports the following operators:

- ReLU;
- ReLU is in place;
- ReLU6;
- ReLU6 is in place;
- ArgMax;
- ArgMin;
- Add Operator;
- Conv2D;
- QuantizedConv2D;
- OptConv2D;
- MinPool;
- MaxPool;
- AvgPool;
- GenericPool;
- DepthwiseSepConv2D;
- QuantizedDepthwiseSepConv2D;
- OptDepthwiseSepConv2D;
- Min;
- Max;

- Squeeze;
- MatMul;
- QuantizedFullyConnected;
- Reshape.

uTensor is a rapidly developing project, very lightweight, based on C++, and very efficient. Plus, it's very flexible and allows users to customize it. If your embedded device has very limited resources, uTensor is a good choice.

8.6   Apache TVM

Apache TVM (Tensor Virtual Machine, tensor virtual machine) (Apache Software Foundation, n.d.) is an end-to-end open-source machine learning compiler framework for CPUs, GPUs, and specialized machine learning accelerators. Its goal is to be able to efficiently optimize and run computations on any hardware backend. It attempts to bridge the gap between productivity-focused deep learning frameworks and performance or efficiency-oriented hardware backends. Its frame diagram is shown in Fig. 8.20.

TVM provides the following main functions:

- Compile deep learning models in Keras, MXNet, PyTorch, Tensorflow, CoreML, DarkNet into minimum deployable modules on various hardware backends.
- Infrastructure for automatically generating and optimizing tensor operators on more backends with better performance.

TVM began as a research project within the SAMPL research group at the Paul G. Allen School of Computer Science and Engineering at the University of Washington. The project is now an incubated effort by the Apache Software Foundation (ASF), driven by an open-source community that engages multiple industry and academic institutions in the Apache way.

Fig. 8.20 Apache TVM framework

TVM is as follows:

There is a growing need to bring machine learning to a variety of hardware devices. Current frameworks rely on vendor-specific operator libraries and are optimized for a narrow range of server-class GPUs. Deploying workloads to new platforms—such as mobile phones, embedded devices, and accelerators (e.g., FPGAs, ASICs)—requires significant manual effort. We propose TVM, a compiler that exposes graph-level and operator-level optimizations to provide performance portability for deep learning workloads across different hardware backends. TVM addresses deep learning-specific optimization challenges such as advanced operator fusion, mapping to arbitrary hardware primitives, and memory latency hiding. It also automatically optimizes low-level programs based on hardware characteristics by employing a novel, learning-based cost modeling approach to quickly explore code optimizations. Experimental results show that TVM provides performance across hardware backends that is comparable to state-of-the-art, hand-tuned libraries for low-power CPUs, mobile GPUs, and server-class GPUs. We also demonstrate the ability of TVM to target new accelerator backends, such as a general-purpose FPGA-based deep learning accelerator. The system is open source and used in production within several large companies.

TVM provides two levels of optimization.

1. Computational graph optimization, performing tasks such as advanced operator fusion, model transformation, and memory management.
2. Tensor operator optimization and code generation layer optimization.

TVM provides two main functions:

1. Compile deep learning models into the smallest deployable module.
2. Provides a framework to automatically generate and optimize models on more backends with better performance.

Key features and capabilities of TVM:

1. Performance: Optimizing compilers and minimal runtimes can often free up machine learning workloads on existing hardware.
2. Run anywhere: Supports CPU, GPU, browser, microcontroller, FPGA, etc. Automatically generate and optimize tensor operators in more backends.
3. Flexibility: Supports block sparsity, quantization (1, 2, 4, 8-bit integers), random forest/classical machine learning, memory planning, MISRA-C compatibility, Python prototyping, or all the above
4. Ease of use: Compile deep learning models of Keras, MXNet, PyTorch, Tensorflow, CoreML, DarkNet, etc. Start using TVM in Python on day 1, and build production software in C++, Rust, or Java on day 2.

The workflow of Chen et al. (2018) Apache TVM is shown in Fig. 8.21.
It consists of several steps.

1. Import. The front-end component imports a model into an IRModule, which contains a collection of functions that internally represent the model.

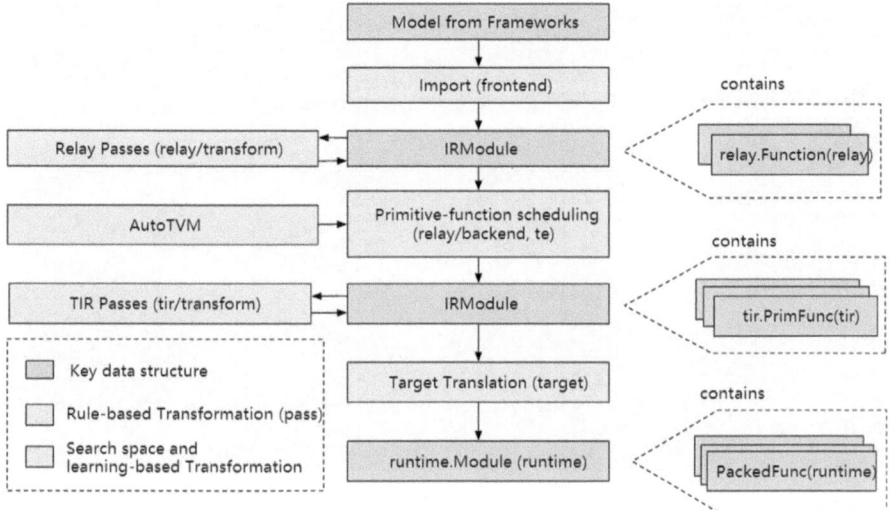

Fig. 8.21 Apache TVM workflow

2. Convert. The compiler converts one IRModule into another IRModule that is functionally equivalent or nearly equivalent (e.g., in the case of quantization). It adopts Tensor-level IR (Tensor Layer Intermediate Representation, TIR for short) format. Many transformations are target (backend) independent. We also allow targets to influence the configuration of the transformation pipeline.
3. Target translation. The compiler translates (codegen) the IRModule into the target-specified executable format. The target translation result is encapsulated as a runtime module that can be exported, loaded, and executed in the target runtime environment.
4. Runtime execution: The user loads back a runtime module and runs the compiled function in a supported runtime environment.

8.7 Qualcomm AI Stack

Qualcomm AI Stack (Qualcomm Technologies, n.d.-b) is a unified AI software portfolio for our mobile, automotive, XR, compute, IoT, and cloud platforms. It is designed to help developers optimize and deploy AI models quickly by supporting AI frameworks and runtimes, developer libraries, system software, and popular operating systems. It is shown in Fig. 8.22.

There are Two SDKs to implement Qualcomm AI Stack.

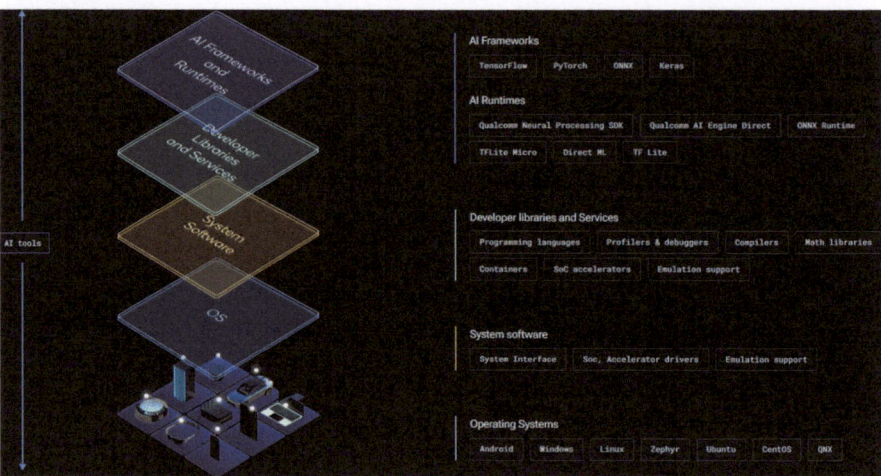

Fig. 8.22 Qualcomm AI Stack

- Qualcomm Neural Processing SDK
- An all-in-one SDK that supports heterogeneous computing and system-level configurations, designed to direct AI workloads to all accelerator cores on our platforms. Provides developers flexibility, including inter-core collaboration support and other advanced features.
- Qualcomm AI Engine Direct SDK
- Provides lower level, highly customizable unified APIs that speeds up AI models on all AI accelerator cores with individual libraries. Can be used directly to target a specific accelerator core or delegate workloads from popular runtimes including Qualcomm Neural Processing SDK, TensorFlow Lite, and ONNX runtime.

8.7.1   Qualcomm Neural Processing SDK

The Qualcomm Neural Processing SDK (Qualcomm Technologies, n.d.-c) is engineered to assist developers in optimizing the performance of trained neural networks on devices equipped with Qualcomm AI products, thereby saving them time and effort. As part of the Qualcomm AI Stack, it facilitates rapid deployment of AI models, allowing them to run entirely on-device.

Qualcomm's products feature extensive AI processing capabilities, enabling the execution of trained neural networks locally without relying on cloud connectivity. The Qualcomm Neural Processing SDK enables developers to run one or multiple neural network models, trained using TensorFlow, PyTorch, Keras, and ONNX, on various Qualcomm platforms, including the CPU, GPU, or Qualcomm Hexagon Processor.

Equipped with tools for model conversion and execution, as well as APIs for targeting specific cores with desired power and performance profiles, the Qualcomm Neural Processing SDK supports convolutional neural networks, custom layers, and more.

By handling much of the computational complexity associated with running neural networks on Qualcomm platforms, the Qualcomm Neural Processing SDK allows developers to devote more time and resources to creating new and innovative user experiences.

The SDK includes

- Android and Linux runtimes for neural network model execution
- Acceleration support for Qualcomm Hexagon™ Processor, Qualcomm Adreno GPUs and Qualcomm Kryo, CPUs
- Support for models in TensorFlow, PyTorch, Keras, and ONNX formats
- APIs for controlling loading, execution, and scheduling on the runtimes
- Desktop tools for model conversion
- Performance benchmark for bottleneck identification
- Sample code and tutorials
- HTML Documentation

In order to simplify the lives of artificial intelligence developers, the Qualcomm Neural Processing SDK refrains from introducing yet another library of network layers. Instead, it empowers developers to design and train their networks using familiar frameworks, with TensorFlow, PyTorch, Keras, and ONNX being supported from the outset. The development workflow is shown in Fig. 8.23:

Upon completing the design and training phases, the model file must be converted into a ".dlc" (Deep Learning Container) file for utilization by the Snapdragon Neural Processing Engine (NPE) runtime. The conversion tool generates conversion

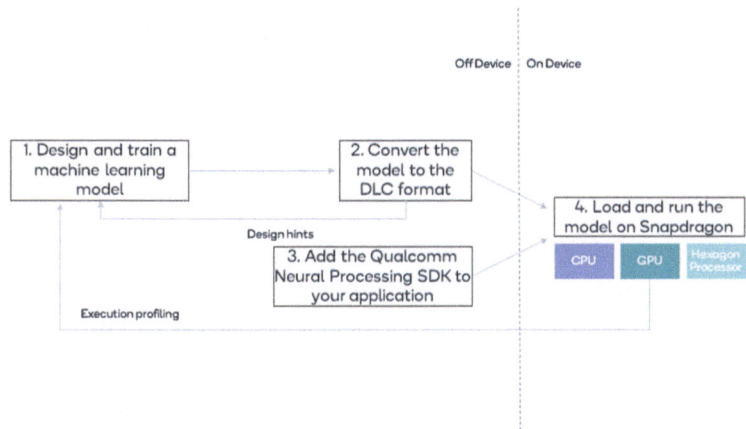

Fig. 8.23 Development workflow

Fig. 8.24 Qualcomm AI Engine Direct SDK

statistics, which include details about unsupported or non-accelerated layers. Developers can utilize this information to refine the design of the initial model.

8.7.2   Qualcomm AI Engine Direct SDK

The Qualcomm AI Engine Direct SDK (Qualcomm Technologies, n.d.-a) offers unified APIs at a lower level for AI development. Developers can enhance the performance of their AI models on Qualcomm AI accelerators—the Qualcomm Kryo CPU, Qualcomm Adreno GPU, and Qualcomm Hexagon processor, by utilizing this SDK. It is shown in Fig. 8.24.

They have the option to target a specific accelerator using the SDK or delegate workloads from TensorFlow Lite or the ONNX runtime to directly access the Hexagon processor.

Features

Qualcomm AI Engine Direct enables a clear separation in software for different hardware cores. The SDK empowers developers to treat Qualcomm AI Engine Direct as a hardware abstraction API, facilitating the porting of applications among various hardware cores. This API provides an appropriate level of abstraction, managing capabilities such as graph optimization internally, while leaving broader functionalities like model parsing and network partitioning to higher-level frameworks.

Developers can adjust the trade-off between capabilities offered by core-specific backend libraries and the associated costs in terms of library size and memory utilization. The resulting compilation produces a high-performance, agile executable with minimal memory footprint, ensuring optimal performance.

Architecture

The architecture is meticulously designed to offer modular, extensible, accelerator-specific libraries, forming a reusable foundation for full-stack AI solutions. It is leveraged by Qualcomm Technologies, Inc. (QTI) within the Qualcomm Neural Processing SDK and serves as a delegator for third-party frameworks like TensorFlow Lite and the ONNX runtime.

The Qualcomm AI Engine Direct architecture furnishes essential components—tools, device, backend, context, graph, operation package registry—for constructing, optimizing, and executing network models on the most suitable core.

8.8   Comparative Analysis of Software Frameworks for Embedded Neural Networks

Some embedded neural network chips and software frameworks have been introduced earlier. They have their own strengths and scope of application. Some software frameworks are customized for specific chips, while others are dedicated to supporting multiple hardware chips. Table 8.1 lists the correspondence between them.

As can be seen from the above table, both TensorFlow Lite and Apache TVM are committed to supporting various types of AI chips, from CPUs and GPUs to dedicated AI accelerators. TensorFlow Lite also supports iPhone and Android phones. Tensor RT, OpenVino, and Vitis are all customized for their own chips.

Table 8.1 Correspondence between embedded AI software framework and AI chip

Embedded AI software framework	Supported AI chips	Supported neural network training frameworks
TensorFlow Lite	Apple, ARM, CPU, GPU	TensorFlow
Tensor RT	NVIDIA GPU (Jetson, Drive, Tesla, etc.)	TensorFlow, Caffe, PyTorch, MXNet
OpenVino	Intel series CPU, GPU, FPGA, VPU	TensorFlow, PyTorch, and MXNet
XILINX Vitis	XILINX DPU (Zynq, Alveo)	TensorFlow, Caffe, PyTorch
uTensor	ARM	Tensorflow
Apache TVM	CPU, GPU, and dedicated machine learning accelerators	Keras, MXNet, PyTorch, TensorFlow, CoreML, DarkNet
Qualcomm AI Stack	Kryo CPU, Adreno GPU, Hexagon DSP	TensorFlow, PyTorch, ONNX, Keras

Various software frameworks are committed to supporting multiple training frameworks, among which Apache TVM has the most comprehensive support. But TensorFlow Lite only supports TensorFlow as a training framework.

At first glance, TensorFlow Lite and Apache TVM support many types of chips, are highly open, and may have broad application prospects. However, since AI chips are computationally intensive, and the inference performance of embedded AI chips is extremely critical, software frameworks optimized for specialized chips are superior in performance. Considering NVIDIA's de facto dominance in the field of AI chips, TensorRT will play a greater role in practical applications. The next "Embedded Artificial Intelligence Implementation" will take TensorRT as an example for in-depth development.

References

Apache Software Foundation. (n.d.) *Apache TVM*. Retrieved from https://tvm.apache.org/

Chen, T., Moreau, T., Jiang, Z., Zheng, L., Yan, E., Cowan, M., Shen, H., Wang, L., Hu, Y., Ceze, L., Guestrin, C., & Krishnamurthy, A. (2018). TVM: An automated end-to-end optimizing compiler for deep learning. arXiv:1802.04799.

NVIDIA Corporation. (2018). *TensorRT integration speeds up TensorFlow inference*. Retrieved from https://devblogs.NVIDIA.com/tensorrt-integration-speeds-tensorflow-inference/

Qualcomm Technologies. (n.d.-a). *Qualcomm AI Engine Direct SDK*. Retrieved from https://developer.qualcomm.com/software/qualcomm-ai-engine-direct-sdk

Qualcomm Technologies. (n.d.-b). *Qualcomm AI Stack*. Retrieved from https://www.qualcomm.com/developer/artificial-intelligence#overview

Qualcomm Technologies. (n.d.-c). *Qualcomm Neural Processing SDK*. Retrieved from https://developer.qualcomm.com/software/qualcomm-neural-processing-sdk

TensorFlow. (n.d.). *TensorFlow Lite | ML for mobile and edge devices*. Retrieved from https://www.tensorflow.org/lite

uTensor. (n.d.). *microTensor*. Retrieved from https://utensor.github.io/website/

XiLinx. (n.d.) *Vitis Platform*. Retrieved from https://www.xilinx.com/products/design-tools/vitis/vitis-platform.html

Part III
Practices

Chapter 9
Embedded AI Development Process

Abstract This chapter introduces the development process of embedded AI applications. First, it summarizes the general development process that can be adapted to various AI acceleration chips, compares its similarities and differences with the general AI application development process, and details the specific development steps for embedded AI development, such as model optimization, conversion, compilation, deployment, etc. Finally, NVIDIA Jetson is taken as an example to introduce its special development process so that developers can gain an intuitive understanding.

Keywords Embedded AI development process · Model optimization

9.1 General Embedded AI Development Process

After understanding the principles of embedded artificial intelligence and becoming familiar with the hardware and software platforms of embedded artificial intelligence, developing embedded AI applications seems to be a matter of course. Then, actual application development is still full of challenges. After all, embedded development is always more complicated than development on general-purpose software and hardware platforms, with more links, more complex processes, and more engineering problems to solve.

The basic workflow of deep neural network development is

1. Prepare the data.
2. Label the data.
3. Build a model.
4. Train the model.
5. Test the model.
6. Improve the model and repeat the process.
7. Save the model for further training or inference.

    For the development of neural network models on embedded devices, several additional steps need to be added based on the above:

© Tsinghua University Press 2024

B. Li, *Embedded Artificial Intelligence*,

https://doi.org/10.1007/978-981-97-5038-2_9

8. Model optimization, which includes training-time optimization, model conversion, compilation, and runtime optimization.

   (a) Optimization during training uses methods such as pruning, and weight sharing to reduce the number of model parameters and compress the size of the model for optimization. It can be seen as a model compression technology.
   (b) Model conversion refers to the process of converting the model from a universal format to the operation set and file format supported by the embedded AI accelerator, such as converting TensorFlow frozen graph to ONNX format. Some model converters also complete the quantization of model parameters to further compress the model.
   (c) Compilation is the process of mapping high-level operations in the model to low-level operation instructions on the embedded AI accelerator, and performing targeted optimization to generate binary code.
   (d) Runtime optimization includes creating a runtime engine and model serialization to optimize the model's use of memory, bandwidth, etc.

9. Model deployment, the converted model must be downloaded to the embedded device for deployment, connect the sensor, obtain real-time data, and run inference.

Among them, steps (3)–(7) are completed on general-purpose AI hardware. This machine is called a development machine or a host machine. It can be a server, PC, or cloud host, etc., and is mainly responsible for the training of AI models. Step 9 is completed on the embedded device, called the inference engine, or the target machine. With an embedded AI accelerator as its core, it is mainly responsible for the inference of AI models and is generally responsible for the collection of field data, such as sensor data, images, video streams, etc. The training-time optimization in step 8 is completed by the development machine, the model conversion and compilation optimization are completed by the development machine for the training machine, and the runtime optimization is completed by the inference engine. Among them, training-time optimization and runtime optimization are optional, as shown in Fig. 9.1.

9.2 NVIDIA Jetson Development Process

A complete development process requires the development machine and the inference engine to be completed collaboratively. During this period, there is a lot of interactive work. In order to assist this workflow, there are some dedicated integrated development environments and tools that can help embedded developers simplify their work, improve efficiency, and focus in the application development itself. Taking the NVIDIA Jetson module as an example, NVIDIA DIGITS (Deep Learning GPU Training System) (NVIDIA Corporation, n.d.-a) software is used to develop complete AI applications.

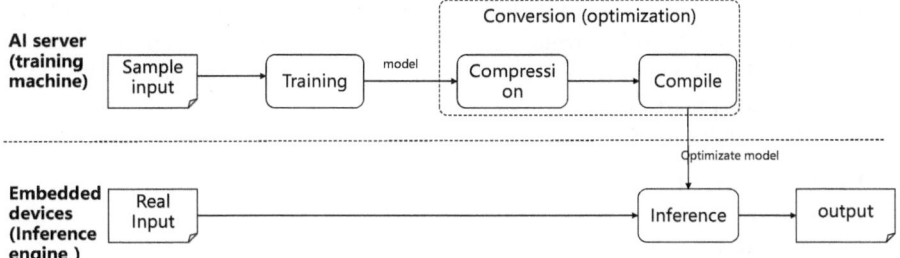

Fig. 9.1 Embedded AI development process

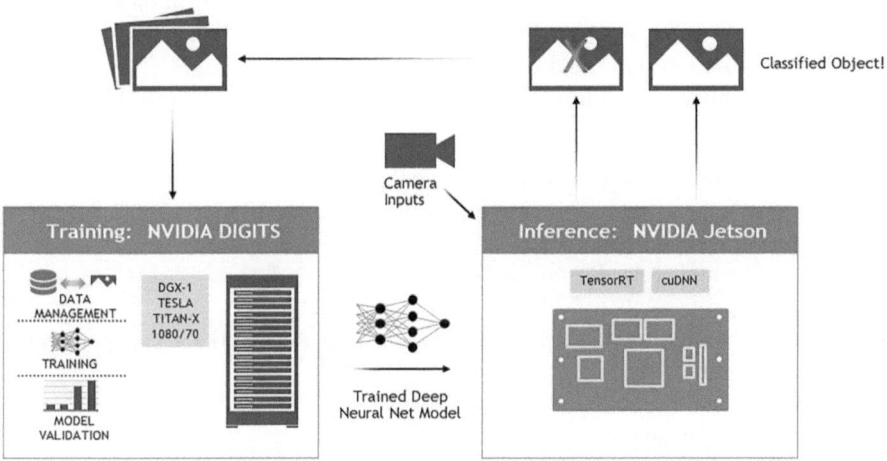

Fig. 9.2 NVIDIA DIGITS development process

As shown in Fig. 9.2, NVIDIA DIGITS runs on the cloud, server, or PC, which is called **a training machine.** It interactively trains the annotated data set to generate a network model. TensorRT and Jetson are deployed in embedded devices, called **inference machines,** for inference at runtime in the field.

DIGITS and TensorRT form an efficient workflow for developing and deploying deep neural networks capable of advanced AI and perception. Among them, DIGITS simplifies common deep learning tasks, such as managing data, designing, and training neural networks on multi-GPU systems, monitoring performance in real time with advanced visualization technology, and selecting the best-performing models from the results browser for deployment. DIGITS is fully interactive, so data scientists can focus on designing and training networks instead of programming and debugging. TensorRT implements model optimization and quantitative inference, providing high-performance deep neural network inference capabilities.

The following takes the NVIDIA Jetson module as an example to build a complete embedded neural network development environment.

First, you need to install the JetPack (Jetson Development Pack) (NVIDIA Corporation, n.d.-b) toolkit, which is an all-in-one software package provided on demand that bundles and installs all development software tools for the NVIDIA Jetson embedded platform.

JetPack includes the following components:

- **Deep learning:** TensorRT, cuDNN, NVIDIA DIGITS.
- **Computer Vision:** NVIDIA Vision Works, OpenCV.
- **GPU computing:** NVIDIA CUDA and library functions.
- **Multimedia:** ISP support, camera images, video CODECs (codecs).

It also includes ROS compatibility, OpenGL, advanced developer tools, and more.

NVIDIA JetPack SDK provides a complete solution for building AI applications. Use the JetPack installer to flash the latest OS image to the Jetson module, install development tools for the host PC and developer suite, and install the libraries, APIs, etc. required to start the development environment.

JetPack installation requires the use of a host machine and mainly includes the following steps:

1. Connect the host computer and Jetson via USB cable and Ethernet cable.
2. Download JetPack to the host machine, which must use Ubuntu operating system.
3. Start the JetPack installer.
4. Select and connect the Jetson series module.
5. Select the components to be installed, including debuggers, analyzers, OpenCV, VisionWorks running on the host, and OpenCV, VisionWorks, DeepStream, etc. on the Jetson module
6. Install the software on the host machine.
7. Flash the OS image onto the module.
8. Install the software on the Jetson module.

NVIDIA now also provides JetPack components in the form of Debian software packages. The JetPack software package can be installed directly on the Jetson module of the flashed machine without the need for a host. Of course, the flashing step must be completed through the host machine.

Finally, TensorRT is installed on the Jetson module and can be used for inference acceleration of neural networks.

References

NVIDIA Corporation. (n.d.-a). *GitHub—NVIDIA/DIGITS: Deep Learning GPU Training System*. Retrieved from https://github.com/NVIDIA/DIGITS
NVIDIA Corporation. (n.d.-b). *NVIDIA JetPack Documentation*. Retrieved from https://docs. nvidia.com/jetson/jetpack/index.html

Chapter 10
Optimizing Embedded Neural Network Models

Abstract This chapter provides an in-depth discussion on model optimization, a key step in embedded AI application development. This includes training-time optimization, post-training optimization, model conversion, model compilation, runtime optimization, and other sub-steps. The pruning, clustering, quantization, compression, and compilation collaboration technologies introduced in the previous chapters are used. In order to deepen readers' understanding, TensorRT, a model optimization tool designed specifically for NVIDIA chips, is introduced in detail.

Keywords Post-training optimization · Training-time optimization · Runtime optimization

The neural network trained on the development machine usually has many parameters and the model size is relatively large, while the memory and external memory on the inference machine (embedded device, such as Jetson module) and the video memory on the embedded AI accelerator are all limited. If you want to install a large neural network into a limited space and run it smoothly, you need to optimize the model, mainly through model compression technology, such as pruning, clustering, quantization, etc. to reduce the size of the model.

10.1 TensorFlow Model Optimization

The TensorFlow Model Optimization Toolkit (Tensorflow, n.d.) is a set of tools that optimize machine learning models for easy deployment and execution. It is part of the TensorFlow Lite project.

The toolkit has many uses, including support for technologies used to:

- Reduce latency and inference costs for cloud and edge devices such as mobile and IoT devices.
- Deploy models to edge devices, which have limitations in processing, memory, power consumption, network connectivity, and model storage space.

© Tsinghua University Press 2024

B. Li, *Embedded Artificial Intelligence*,

https://doi.org/10.1007/978-981-97-5038-2_10

• Execute models and optimize them on existing hardware or new purpose-built accelerators.

The TensorFlow model optimization toolkit can start with ready-made models or models trained by developers themselves for optimization. Models trained by developers themselves can also be optimized during training. Finally, a compressed model file with the suffix name Tflite is generated.

10.1.1   Post-training Optimization

Post-training optimization of TensorFlow models includes general techniques for reducing CPU and hardware accelerator latency, processing time, power consumption, and model size with little loss of model accuracy. These techniques can be performed on already trained floating-point TensorFlow models and applied during TensorFlow Lite conversion. These technologies are optionally enabled in the TensorFlow Lite converter. It contains two methods:

Quantitative Weight

This method converts the weights to a lower precision type, such as a 16-bit floating-point number or an 8-bit integer. It is generally recommended to use 16-bit floating-point numbers for GPU acceleration and 8-bit integers for CPU execution.

An example procedure for specifying the weighting of an 8-bit integer is given below:

```
import tensorflow as tf
converter = tf.lite.TFLiteConverter.from_saved_model
( saved_model_dir )
converter.optimizations = [ tf.lite.Optimize.OPTIMIZE_
FOR_SIZE ]
tflite_quant_model = converter.convert ()
```

In this way, the most critical intensive parts of the inference are computed in 8 bits instead of floating point. Compared with the method of quantizing both weights and activation values introduced below, the performance overhead during inference is slightly larger.

Full Integer Quantization of Weights and Activation Values

This approach quantizes both weights and activations as integers, improving latency, processing, and energy consumption, and gaining access to pure integer hardware accelerators. Implementing this approach requires a small representative data set.

An example procedure for such quantization is given below:

```
import tensorflow as tf

def representative_dataset_gen():
  for _ in range(num_calibration_steps):
    # Get sample input data as a numpy array in a method of
    your choosing.
    yield [input]

converter = tf.lite.TFLiteConverter.from_saved_model(saved_
model_dir)
converter.optimizations = [tf.lite.Optimize.DEFAULT]
converter.representative_dataset = representative_dataset_gen
tflite_quant_model = converter.convert()
```

For convenience, the generated model will still take floating-point numbers as input and output.

After TensorFlow training, the quantization tool completes the quantization (model optimization work) and the conversion of the model (that is, converting from TensorFlow format to TFLite format).

10.1.2 Optimization During Training

When the post-training optimization tool cannot meet the needs, you can further try the training-time optimization tool.

During training, the tool is equipped with the loss function of the model to train the data so that the model can "adapt" to the changes brought about by the optimization technology.

These training-time tools include weight pruning, quantization-aware training, weight clustering, and co-optimization. They are explained separately below.

Weight Pruning

During the training process, magnitude-based weight pruning gradually returns the model weights to zero to achieve model sparsity. Sparse models are easier to compress, and zero values can be skipped during inference to improve latency.

This technique brings 6× improvement through model compression with little accuracy loss.

The following is an end-to-end example of magnitude-based weight pruning. It includes the following parts:

1. Train a tf.keras model from scratch for the MNIST dataset.
2. Fine-tune the model by applying the pruning API and view the accuracy.
3. Create 3× smaller TF and TFLite models through pruning.
4. 10× smaller TFLite model by combining pruning and post-training quantization.
5. See if the accuracy is maintained from TF to TFLite.

First install the TensorFlow optimization toolkit:

```
pip install -q tensorflow-model-optimization
```

Then train a model for the MNIST dataset:

```
import tempfile
import os

import tensorflow as tf
import numpy as np

from tensorflow import keras

% load_ext tensorboard
# Load MNIST dataset
mnist = keras.datasets.mnist
(train_images, train_labels), (test_images, test_labels) =
mnist.load_data()

# Normalize the input image so that each pixel value is
between 0 and 1.
train_images = train_images / 255.0
test_images = test_images / 255.0
```

```
# Define the model architecture.
model = keras.Sequential([
  keras.layers.InputLayer(input_shape=(28, 28)),
  keras.layers.Reshape(target_shape=(28, 28, 1)),
  keras.layers.Conv2D(filters=12, kernel_size=(3, 3),
  activation='relu'),
  keras.layers.MaxPooling2D(pool_size=(2, 2)),
  keras.layers.Flatten(),
  keras.layers.Dense(10)
])

# Train the digit classification model
model.compile(optimizer='adam', loss=tf.keras.losses.
SparseCategoricalCrossentropy(from_logits=True),
metrics=['accuracy'])

model.fit(
  train_images,
  train_labels,
  epochs=4,
  validation_split=0.1,
)
```

Evaluate its accuracy as a baseline and save the model for later use.

```
_, baseline_model_accuracy = model.evaluate(
    test_images, test_labels, verbose=0)

print('Baseline test accuracy:', baseline_model_accuracy)

_, keras_file = tempfile.mkstemp('.h5')
tf.keras.models.save_model(model, keras_file, include_
optimizer=False)
print('Saved baseline model to:', keras_file)
```

The accuracy is 0.9775999784469604.

Next, the pruning method is used to fine-tune the pre-trained model. The entire model will be pruned, starting with 50% sparsity (that is, 50% of the weights are zero) and ending with 80% sparsity.

```
import tensorflow_model_optimization as tfmot

prune_low_magnitude = tfmot.sparsity.keras.prune_low_magnitude

#Calculate the number of steps required to complete 2 epochs
of pruning
batch_size = 128
epochs = 2
validation_split = 0.1 # 10% of training set will be used
for validation set.

num_images = train_images.shape [0] * (1 - validation_split )
end_step = np.ceil ( num_images / batch_size ). astype (np.
int32) * epochs

#Define the model to be pruned
pruning_params = {
' pruning_schedule ': tfmot.sparsity. keras.PolynomialDecay
( initial_sparsity =0.50, final_sparsity =0.80, begin_step =0,
end_step = end_step )
}

model_for_pruning = prune_low_magnitude (model, ** pruning_
params )

#Recompile the model for pruning
model_for_pruning.compile (optimizer='adam',
loss=tf.keras.losses.SparseCategoricalCrossentropy(from_
logits=True), metrics=[ 'accuracy'])

model_for_pruning.summary ()
```

Run the above program and see that about 50 % of the weights can be pruned.

```
Total params: 40,805
Trainable params: 20,410
Non-trainable params: 20,395
```

Use pruning to fine-tune two epochs and evaluate the results.

```
logdir = tempfile.mkdtemp()

callbacks = [ tfmot.sparsity.keras.UpdatePruningStep(),
  tfmot.sparsity.keras.PruningSummaries(log_dir=logdir),
]

model_for_pruning.fit(train_images, train_labels, batch_
size=batch_size, epochs=epochs, validation_split=validation_
split, callbacks=callbacks)
_, model_for_pruning_accuracy = model_for_pruning.evaluate
(test_images , test_labels , verbose=0)

print('Baseline test accuracy:', baseline_model_accuracy )
print('Pruned test accuracy:', model_for_pruning_accuracy )
```

It can be seen that the accuracy is basically maintained and only dropped by 0.005.

```
Baseline test accuracy: 0.9775999784469604
Pruned test accuracy: 0.972100019454956
```

Next, create a model reduced by three times by trimming. Two techniques are used here, tfmot.sparsity.keras.strip_pruning and standard compression algorithms (e.g., via gzip). strip_pruning removes tf.Variable which is only needed during training and would otherwise increase model size during inference. Standard compression algorithms are also necessary because the serialized weight matrix is the same size as it was before pruning. However, pruning makes most of the weights zero, and this additional redundancy can be further compressed using algorithms.
Code shows as below:

```
# Create compressible models for TensorFlow
model_for_export = tfmot.sparsity.keras.strip_pruning ( model_
for_pruning )

_, pruned_keras_file = tempfile.mkstemp ('.h5')
tf.keras.models.save_model ( model_for_export , pruned_
keras_file , include_optimizer =False)
print('Saved pruned Keras model to: ', pruned_keras_file )
# Create compressible models for TensorFlow Lite
converter = tf.lite.TFLiteConverter.from_keras_model(model_
for_export)
pruned_tflite_model = converter.convert()
```

```
_, pruned_tflite_file = tempfile.mkstemp('.tflite')

with open(pruned_tflite_file, 'wb') as f:
  f.write(pruned_tflite_model)

print('Saved pruned TFLite model to:', pruned_tflite_file)
# Define a helper function to compress the model via gzip and
measure the compressed size
def get_gzipped_model_size(file):
  # Returns size of gzipped model, in bytes.
  import os
  import zipfile

  _, zipped_file = tempfile.mkstemp('.zip')
  with zipfile.ZipFile(zipped_file, 'w', compression=zipfile.
ZIP_DEFLATED) as f:
    f.write(file)

  return os.path.getsize(zipped_file)
# Show model size
print("Size of gzipped baseline Keras model: %.2f bytes" %
(get_gzipped_model_size(keras_file)))
print("Size of gzipped pruned Keras model: %.2f bytes" %
(get_gzipped_model_size(pruned_keras_file)))
print("Size of gzipped pruned TFlite model: %.2f bytes" %
(get_gzipped_model_size(pruned_tflite_file)))
```

Comparing the output results, we found that the model is three times smaller than before pruning.

```
Size of gzipped baseline Keras model: 78211.00 bytes
Size of gzipped pruned Keras model: 25797.00 bytes
Size of gzipped pruned TFlite model: 24995.00 bytes
```

If post-training quantization is added to the pruned model, the model can be compressed ten times.

```
converter = tf.lite.TFLiteConverter.from_keras_model(model_
for_export)
converter.optimizations = [tf.lite.Optimize.DEFAULT]
quantized_and_pruned_tflite_model = converter.convert()
```

```
_, quantized_and_pruned_tflite_file = tempfile.mkstemp('.tflite')

with open(quantized_and_pruned_tflite_file, 'wb') as f:
  f.write(quantized_and_pruned_tflite_model)

print('
Saved quantized and pruned TFLite model to:',
quantized_and_pruned_tflite_file)

print("Size of gzipped baseline Keras model: %.2f bytes" %
(get_gzipped_model_size(keras_file)))
print("Size of gzipped pruned and quantized TFlite model:
%.2f bytes" %
(get_gzipped_model_size(quantized_and_pruned_tflite_file)))
```

As you can see, the model is compressed from 78 to 8 KB size.

```
Size of gzipped baseline Keras model: 78211.00 bytes
Size of gzipped pruned and quantized TFlite model: 8031.00 bytes
```

Finally, observe whether the accuracy of the model is maintained before and after compression.

```
# Define a helper function to evaluate the TF Lite model on
the test data set
import numpy as np

def evaluate_model(interpreter):
  input_index = interpreter.get_input_details()[0]["index"]
  output_index = interpreter.get_output_details()[0]["index"]

  # Run predictions on ever y image in the "test" dataset.
prediction_digits = []
  for i, test_image in enumerate(test_images):
    if i % 1000 == 0:
      print('Evaluated on {n} results so far.'.format(n=i))
    # Pre-processing: add batch dimension and convert to float32
to match with the model's input data format.
    test_image = np.expand_dims(test_image, axis=0).
astype(np.float32)
```

```
    interpreter.set_tensor(input_index, test_image)

    # Run inference.    interpreter.invoke()

    # Post-processing: remove batch dimension and find the
digit with highest probability.
    output = interpreter.tensor(output_index)
    digit = np.argmax(output()[0])
    prediction_digits.append(digit)

  print('\n')
# Compare prediction results with ground truth labels to
calculate accuracy.
  prediction_digits = np.array(prediction_digits)
  accuracy = (prediction_digits == test_labels).mean()
  return accuracy
# Evaluate before and after cropping
interpreter =
tf.lite.Interpreter(model_content=quantized_and_pruned_
tflite_model)
interpreter.allocate_tensors()

test_accuracy = evaluate_model(interpreter)

print('Pruned and quantized TFLite test_accuracy:', test_
accuracy)
print('Pruned TF test accuracy:', model_for_pruning_accuracy)
print('Baseline test accuracy:', baseline_model_accuracy)
```

Finally, it was found that after applying pruning and quantization, its accuracy was maintained (or even slightly increased) compared to the pruned model during training and was only slightly lower than the original model before pruning (about 0.4%).

```
Pruned and quantized TFLite test_accuracy : 0.9722
Pruned TF test accuracy: 0.972100019454956
Baseline test accuracy: 0.9775999784469604
```

Quantitative Perception Training

Quantization comes in two forms: post-training quantization and quantization-aware training. Post-training quantization is easier to use, but quantization-aware training generally maintains model accuracy better.

Quantization-aware training simulates quantization at inference time, creating a model that will be used by downstream tools to produce the actual quantization model. Quantized models use lower precision (e.g., 8-bit instead of 32-bit floating point), which provides benefits during deployment.

Quantization brings improvements through model compression and latency reduction. Using API defaults, model size can be reduced by 4 times and CPU latency can be reduced by 1.5–4 times. Eventually further latency improvements will be achieved in compatible embedded AI accelerators such as Google Edge TPU.

Quantitative awareness training has the following limitations:

1. Model: Model only uses permissionlisted layers, BatchNormalization (when it follows Conv2D and DepthwiseConv2D layers), and in limited cases Concat.
2. Hardware acceleration: Default with EdgeTPU, Android NNAPI (Neural Network API), and TFLite backend, etc. compatible.
3. Quantization deployment: Currently only per-axis quantization of convolutional layers is supported, not per-tensor quantization.

The following is an end-to-end example of quantization-aware training. It includes the following parts:

1. Train a tf.keras model for MNIST from scratch.
2. Fine-tune the model, view accuracy, and export the quantization-aware model by applying the quantization-aware training API.
3. Use this model to create a realistic quantized model for the TFLite backend.
4. Observe whether accuracy is maintained.

The first part will not be described again. Let's start by training an existing model using quantization awareness. Please note that the model generated in this step is quantization aware but does not complete quantization (for example, the weights are float32 instead of INT8).

First, a quantization-aware training model is created.

```
import tensorflow_model_optimization as tfmot

quantize_model = tfmot.quantization.keras.quantize_model

q_aware_model = quantize_model ( model) # model is an
existing model

#Recompile q_aware_model.compile
(optimizer=' adam ',
loss=tf.keras.losses.SparseCategoricalCrossentropy(from_
logits =True),
metrics=['accuracy'])

q_aware_model.summary ()
```

Then, the model is fine-tuned using quantization-aware training on a subset of the training data, training for just one epoch and observing changes in accuracy before and after training.

```
train_images_subset = train_images[0:1000] # out of 60000
train_labels_subset = train_labels[0:1000]

q_aware_model.fit(train_images_subset, train_labels_subset,
               batch_size=500, epochs=1, validation_split=0.1)
_, baseline_model_accuracy = model.evaluate (
   test_images , test_labels , verbose=0)

_, q_aware_model_accuracy = q_aware_model.evaluate (
   test_images , test_labels , verbose=0)

print('Baseline test accuracy:', baseline_model_accuracy )
print('Quant test accuracy:', q_aware_model_accuracy )
```

It was observed that the accuracy was maintained or even increased slightly.

```
Baseline test accuracy: 0.9609000086784363
Quant test accuracy: 0.9628999829292297
```

Now convert to a quantized model in TensorFlow Lite format.

```
converter = tf.lite.TFLiteConverter.from_keras_model ( q_
aware_model )
converter.optimizations = [ tf.lite.Optimize.DEFAULT ]

quantized_tflite_model = converter.convert ()
```

This results in an actual quantized model with INT8 weights and UINT8 activations.

Finally, compare the accuracy and size changes before and after model optimization. The specific code will not be described again, the results are shown below.

```
Quant TFLite test_accuracy : 0.963
Quant TF test accuracy: 0.9628999829292297
Baseline test accuracy: 0.9609000086784363
Float model in Mb: 0.08053970336914062
Quantized model in Mb: 0.02339935302734375
```

The accuracy of the model is maintained, while the size is reduced by four times.

Weight Clustering

Weight clustering, or weight sharing, reduces the number of duplicates of unique weight values in the model, thereby surrogate deployment benefits. It first divides the weights of each layer into N groups, and then all the weights belonging to this group share the centroid value of the group.

This method is used for tasks such as vision and speech, and the model compression rate can be increased by five times with a small loss in accuracy.

An end-to-end example of weight clustering is given below, including the following parts:

1. Train a tf.keras model from scratch for the MNIST dataset.
2. Fine-tune the model by applying the weighted clustering API and view the accuracy.
3. Create a 6× smaller model of TF and TFLite via clustering.
4. Create an 8× smaller TFLite model by combining weight clustering and post-training quantization.
5. Observe whether the accuracy is maintained from TF to TFLite.

Part 1 will not be described again. The code for Part 2 is as follows.

```
import tensorflow_model_optimization as tfmot

    cluster_weights = tfmot.clustering.keras.cluster_weights
    CentroidInitialization = tfmot.clustering.keras.
CentroidInitialization

    clustering_params = {' number_of_clusters ': 16, '
cluster_centroids_init   ': CentroidInitialization.LINEAR
    }

#Cluster the entire model
clustered_model = cluster_weights ( model, ** clustering_
params )

#Use a smaller learning rate to fine-tune the clustering model
opt = tf.keras.optimizers.Adam ( learning_rate =1e-5)

clustered_model.compile (
loss=tf.keras.losses.SparseCategoricalCrossentropy(from_
logits= True), optimizer=opt, metrics=['accuracy'])
```

```
clustered_model.summary ()
# Fine-tune the model, only need to run  for one epoch
clustered_model.fit
  train_images ,
  train_labels ,
  batch_size =500,
  epochs=1,
  validation_split =0.1)
# Save model
final_model = tfmot.clustering.keras.strip_clustering
(clustered_model)

_, clustered_keras_file = tempfile.mkstemp('
.h5')print('Saving clustered model to: ', clustered_keras_file)
tf.keras.models.save_model(final_model, clustered_keras_file,
                        include_optimizer=False)
# The compressed model is in tflite format
clustered_tflite_file = '/tmp/clustered_mnist.tflite'
converter = tf.lite.TFLiteConverter.from_keras_model(final_
model)
tflite_clustered_model = converter.convert()
with open(clustered_tflite_file, 'wb') as f:
  f.write(tflite_clustered_model)
print('Saved clustered TFLite model to:', clustered_tflite_file)
print("Size of gzipped baseline Keras model: %.2f bytes" %
( get_gzipped_model_size ( keras_file )))
print("Size of gzipped clustered Keras model: %.2f bytes" %
( get_gzipped_model_size ( clustered_keras_file )))
print("Size of gzipped clustered TFlite model: %.2f bytes" %
( get_gzipped_model_size ( clustered_tflite_file )))
```

It can be observed that the model size is reduced by six times.

```
Size of gzipped baseline Keras model: 78047.00 bytes
Size of gzipped clustered Keras model: 12524.00 bytes
Size of gzipped clustered TFlite model: 12141.00 bytes
```

Combined with post-training quantization:

```
converter = tf.lite.TFLiteConverter.from_keras_model(final_
model)
converter.optimizations = [tf.lite.Optimize.DEFAULT]
tflite_quant_model = converter.convert()

_, quantized_and_clustered_tflite_file = tempfile.mkstemp('.
tflite')

with open(quantized_and_clustered_tflite_file, 'wb') as f:
  f.write(tflite_quant_model)
```

Model size can be further reduced while accuracy is still maintained:

```
Size of gzipped clustered and quantized TFlite model:
9240.00 bytes
Clustered and quantized TFLite test_accuracy : 0.9746
Clustered TF test accuracy: 0.9746000170707703
Baseline test accuracy: 0.9785000085830688
```

Collaborative Optimization

Collaborative optimization is an overarching process across a variety of techniques that produces models that, when deployed, achieve the best balance of target characteristics such as inference speed, model size, and accuracy.

It is not difficult to imagine that collaborative optimization uses a combination of the individual technologies introduced above to achieve a comprehensive optimization effect. Various combinations are possible.

But the problem that arises when trying to combine these techniques is that applying one technique often destroys the results of the previous technique, thus defeating the overall advantage of applying all of them simultaneously. For example, clustering techniques do not preserve the sparsity introduced by pruning techniques. To solve this problem, the following collaborative optimization technology is introduced:

1. Sparsity preserving clustering
2. Sparsity preserving quantization-aware training (PQAT)
3. Cluster preserving quantization-aware training (CQAT)
4. Sparsity and cluster preserving quantization-aware training (PCQAT)

This provides multiple deployment paths for compressing machine learning models and leveraging hardware acceleration at inference time. Figure 10.1 shows

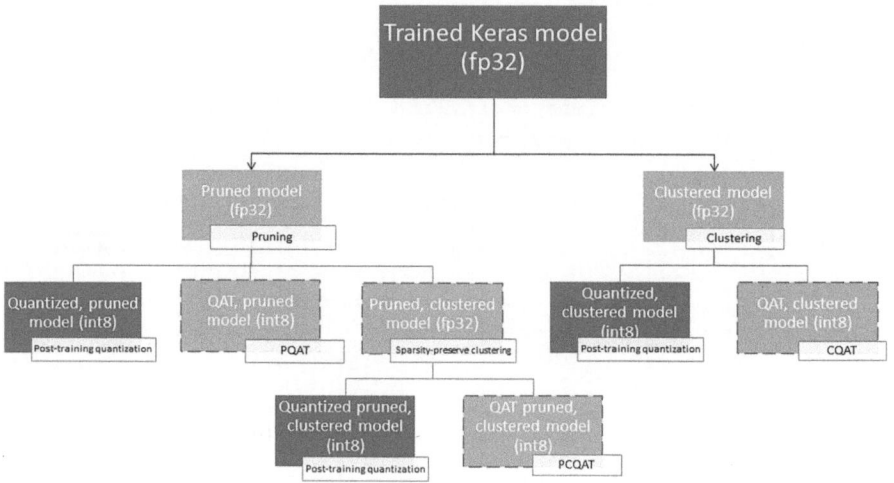

Fig. 10.1 TensorFlow Lite collaborative optimization deployment path

the main deployment paths, one can choose a certain path to obtain a model with the required deployment characteristics, where the leaf nodes are deployable models, which means that they are partially or fully quantized and in TFLite format. Light fills indicate steps requiring retraining/fine-tuning, and dashed borders highlight co-optimization steps. The technology used on a given node is indicated in its label. Among them, QAT is the abbreviation of quantization-aware training.

A fully optimized model will be reached at the third level of the above deployment tree. However, any other level of optimization may prove satisfactory and achieve the desired balance of inference latency and accuracy, in which case no further optimization is required. The recommended training process is to iterate through all nodes applicable to the target deployment scenario to see if the model meets the inference latency requirements. If not, use the corresponding collaborative optimization technology to further compress the model, and repeat this process until the model is fully optimized (pruning, clustering, and quantization).

Figure 10.2 shows the changes in the density map of sample weights when passing through the collaborative optimization pipeline.

The result is a quantized deployment model with reduced unique values and a large number of sparse weights, depending on the target sparsity specified at training time. In addition to significant model compression benefits, specific hardware support can leverage these sparse, aggregated models to significantly reduce inference latency.

Taking CQAT (Cluster Preserving Quantitative Aware Training) as an example, an end-to-end example program is given below, including the following parts:

1. Train tf.keras model from scratch for the MNIST dataset.
2. Fine-tune the model by applying the weighted clustering API and view the accuracy.

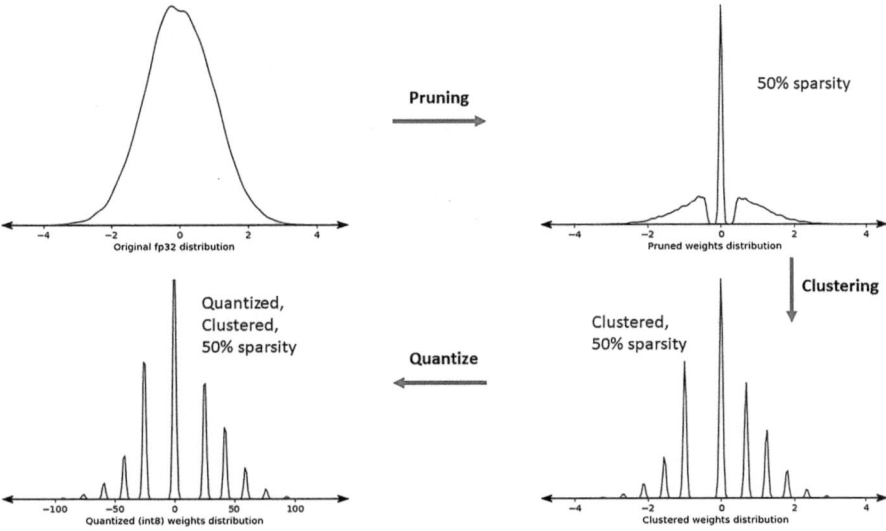

Fig. 10.2 Density map changes of sample weights during TensorFlow Lite collaborative optimization pipeline processing

3. Apply quantization-aware training and observe that clusters are lost.
4. Apply distance-preserving quantization-aware training and observe that previously applied clustering is preserved.
5. Generate a TFLite model and observe the effect of applying CQAT to it.
6. The obtained CQAT model accuracy was compared with the accuracy of the model using post-training quantization.

Among them, Part 1 will not be described again. The core of Part 2 is to apply the cluster_weights () API to cluster the entire pre-trained model. The code shows as below:

```
import tensorflow_model_optimization as tfmot
cluster_weights = tfmot.clustering.keras.cluster_weights
CentroidInitialization = tfmot.clustering.keras.
CentroidInitialization
clustering_params = {
' number_of_clusters ': 8,
  'cluster_centroids_init': CentroidInitialization.KMEANS_
PLUS_PLUS
}
clustered_model = cluster_weights(model, **clustering_params)
# Fine-tune with a smaller learning rate
opt = tf.keras.optimizers.Adam(learning_rate=1e-5)
```

```
clustered_model.compile(
  loss=tf.keras.losses.SparseCategoricalCrossentropy(from_
logits=True),
  optimizer=opt,
  metrics=['accuracy'])
clustered_model.summary()
# Fine-tuning the clustering model for 3 epochs
clustered_model.fit (
  train_images ,
  train_labels ,
epochs=3,
  validation_split =0.1)
# Define a helper function to count and print the number of
clusters in each kernel of the model.
def print_model_weight_clusters (model):
for layer in model.layers :
if isinstance (layer, tf.keras.layers.Wrapper ):
            weights = layer.trainable_weights
        else:
            weights = layer.weights
        for weight in weights:
            # ignore auxiliary quantization weights
            if "quantize_layer" in weight.name:
                continue
            if "kernel" in weight.name:
                unique_count = len(np.unique(weight))
                print(
                    f"{layer.name}/{weight.name}: {unique_
count} clusters "
)
# To check whether the model kernel is clustered correctly.
We need to peel off the clustering wrapper first.
stripped_clustered_model = tfmot.clustering.keras.strip_
clustering ( clustered_model )
print_model_weight_clusters ( stripped_clustered_model )
_, clustered_model_accuracy = clustered_model.evaluate (
  test_images , test_labels , verbose=0)
print('Baseline test accuracy:', baseline_model_accuracy )
print('Clustered test accuracy:', clustered_model_accuracy )
```

The model kernel is clustered into eight clusters with almost no loss in accuracy.

```
conv2d/kernel:0: 8 clusters
dense/kernel:0: 8 clusters
Baseline test accuracy: 0.9814000129699707
Clustered test accuracy: 0.9800999760627747
```

Next, we apply quantization-aware training (QAT) and cluster preserving quantization-aware training (CQAT) on the clustering model and observe that CQAT preserves weight clustering in the clustering model. Note that before applying the CQAT API, use tfmot.clustering.keras.strip_clustering to strip the clustering wrapper from the model.

```python
# QAT , quantified awareness training
qat_model =
tfmot.quantization.keras.quantize_model(stripped_clustered_
model)
qat_model.compile (optimizer=' adam ',
loss=tf.keras.losses.SparseCategoricalCrossentropy(from_
logits=True),
metrics=['accuracy'])
print('Train qat model:')
qat_model.fit ( train_images , train_labels , batch_size =128,
epochs=1, validation_split =0.1)
# CQAT , cluster-preserving quantization-aware training
quant_aware_annotate_model =
tfmot.quantization.keras.quantize_annotate_model(
                stripped_clustered_model)
cqat_model = tfmot.quantization.keras.quantize_apply(
                quant_aware_annotate_model,
tfmot.experimental.combine.
Default8BitClusterPreserveQuantizeScheme())
cqat_model.compile(optimizer='adam',

loss=tf.keras.losses.SparseCategoricalCrossentropy(from_
logits=True),
                metrics=['accuracy'])
print('Train cqat model:')
cqat_model.fit ( train_images , train_labels , batch_size =128,
epochs=1, validation_split =0.1)
```

Observe the clustering effect of the two:

```
print("QAT Model clusters:")
print_model_weight_clusters ( qat_model )
print("CQAT Model clusters:")
print_model_weight_clusters ( cqat_model )
```

It can be seen that the former loses the clustering effect, while the latter maintains the clustering effect.

```
QAT Model clusters:
quant_conv2d/conv2d/kernel:0: 108 clusters
quant_dense /dense/kernel:0: 19931 clusters
CQAT Model clusters:
quant_conv2d/conv2d/kernel:0: 8 clusters
quant_dense /dense/kernel:0: 8 clusters
```

Next, the model is converted to TFLite and the TFLite model is evaluated to ensure accuracy is maintained.

```
# Define helper functions to obtain compressed model files.
def get_gzipped_model_size (file):
# It returns the size of the gzipped model in kilobytes.
  _, zipped_file = tempfile.mkstemp('.zip')
  with zipfile.ZipFile(zipped_file, 'w', compression=zipfile.ZIP_
DEFLATED) as f:
    f.write(file)
  return os.path.getsize(zipped_file)/1000
# Convert the QAT model.
converter = tf.lite.TFLiteConverter.from_keras_model(qat_model)
converter.optimizations = [tf.lite.Optimize.DEFAULT]
qat_tflite_model = converter.convert()
qat_model_file = 'qat_model.tflite'
# Save the model.
with open(qat_model_file, 'wb') as f:
    f.write(qat_tflite_model)
# Convert the CQAT model.
converter = tf.lite.TFLiteConverter.from_keras_model(cqat_model)
converter.optimizations = [tf.lite.Optimize.DEFAULT]
cqat_tflite_model = converter.convert()
cqat_model_file = 'cqat_model.tflite'
```

```
# Save the model.
with open(cqat_model_file, 'wb') as f:
    f.write(cqat_tflite_model)
print("QAT model size: ", get_gzipped_model_size(qat_model_
file), ' KB')
print("CQAT model size: ", get_gzipped_model_size ( cqat_
model_file ), ' KB')
# Define a helper function to evaluate the TFLite model on
the test data set .
def eval_model (interpreter):
  input_index = interpreter.get_input_details ()[0]["index"]
  output_index = interpreter.get_output_details ()[0]["index"]
#Predict each image on the test data set
  prediction_digits = []
for i , test_image in enumerate( test_images ):
if i % 1000 == 0:
print( f"Evaluated on { i } results so far.")
#Preprocessing : Increase the dimensions of the batch and
convert the data to 3 2 -bit floating point numbers to match
the input data format of the model
    test_image = np.expand_dims ( test_image , axis=0).
astype (np.float32)
    interpreter.set_tensor ( input_index , test_image )
#Run inference
    interpreter.invoke ()
# Post-processing: reduce the dimension of the batch, and
then find the number with the highest confidence
output = interpreter.tensor ( output_index )
digit = np.argmax (output()[0])
    prediction_digits.append (digit)
print('\n')
#Compare the prediction results with the real annotations
and calculate the accuracy
  prediction_digits = np.array(prediction_digits)
  accuracy = (prediction_digits == test_labels).mean()
  return accuracy
# Evaluate the accuracy of the CQAT model.
interpreter = tf.lite.Interpreter(cqat_model_file)
interpreter.allocate_tensors()
cqat_test_accuracy = eval_model(interpreter)
print('Clustered and quantized TFLite test_accuracy :',
cqat_test_accuracy )
print('Clustered TF test accuracy:', clustered_model_
accuracy )
```

The size of the CQAT model is further reduced compared to the QAT model, while the accuracy is maintained.

```
QAT model size: 16.685 KB
CQAT model size: 10.121 KB
Clustered and quantized TFLite test_accuracy : 0.9795
Clustered TF test accuracy: 0.9800999760627747
```

Finally, post-training quantization is directly used on the clustering model, and its accuracy is compared with the CQAT model.

```
# Define a generator for the calibration dataset from the
first 1000 training images.
def mnist_representative_data_gen ():
for image in train_images [:1000]:
image = np.expand_dims (image, axis=0). astype (np.float32)
yield [image]
# Generate post-training quantized model from original model
converter =
tf.lite.TFLiteConverter.from_keras_model(stripped_clustered_
model)
converter.optimizations = [tf.lite.Optimize.DEFAULT]
converter.representative_dataset = mnist_representative_
data_gen
post_training_tflite_model = converter.convert()
post_training_model_file = 'post_training_model.tflite'
# Save the model.
with open(post_training_model_file, 'wb') as f:
    f.write(post_training_tflite_model)
# Compare accuracy.
interpreter = tf.lite.Interpreter ( post_training_model_file )
interpreter.allocate_tensors ()
post_training_test_accuracy = eval_model (interpreter)
print('CQAT TFLite test_accuracy :', cqat_test_accuracy )
print('Post-training (no fine-tuning) TF test accuracy:',
post_training_test_accuracy )
```

It can be seen that the accuracy of CQAT is slightly lower because the quantization model has been fine-tuned after training.

Through the above example, we see that the benefit of collaborative optimization is to maintain the results of another optimization when applying one optimization, making the final model smaller in size, and maintaining the accuracy of the model to the greatest extent. The other two collaborative optimizations—pruning-preserving quantization-aware training and sparse-preserving clustering are similar to this and will not be described again.

In short, TensorFlow has a rich set of model optimization tools, including pruning, clustering, quantization , etc. Quantization is divided into training-time quantization and post-training quantization , and multiple model optimization technologies can be used in combination to form a collaborative optimization effect. Developers can choose appropriate optimization techniques according to their own needs and find the best balance between model size and accuracy.

10.2 TensorRT Model Optimization

Different from TensorFlow Lite, TensorRT is a model optimization tool specially implemented for NVIDIA GPU. It can not only import TensorFlow models, but also models generated by training frameworks such as PyTorch and MXNET. Of course, because it is an inference engine, it does not include training-time optimization tools. Instead, it completes model conversion, quantization, compilation optimization, and runtime optimization for specific hardware after training.

10.2.1 Integration with Mainstream Deep Learning Frameworks

TensorRT can be integrated with the following mainstream deep learning frameworks to optimize inference performance. If the model you trained is in the ONNX format or other popular frameworks such as TensorFlow and MATLAB, there is an easy way to import the model into TensorRT for inference. Below are some frameworks listed.

1. TensorFlow
   TensorRT and TensorFlow are tightly integrated, combining the powerful optimization capabilities of TensorRT with the flexibility of TensorFlow.
2. MATLAB
   TensorRT through GPU Coder to automatically generate high-performance inference engines for NVIDIA Jetson, DRIVE and Tesla platforms.
3. ONNX
   TensorRT provides an ONNX parser, so ONNX models can be easily imported into TensorRT from frameworks such as PyTorch, MxNet, Caffe 2, Chainer, Microsoft Cognitive Toolkit, etc. TensorRT can also be integrated with the ONNX Runtime, providing an easy way to achieve high-performance inference of machine learning models in the ONNX format.

Integrate with TensorFlow

Overview

TensorRT can be used with TensorFlow in the following ways (NVIDIA Corporation, 2018).

1. UFF (Universal Framework Format) method: This method is only effective if the entire graph can be converted to UFF (Universal Framework Format) and can be accelerated by TensorRT.
2. TF-TRT (TensorFlow-TensorRT) method: This method integrates TensorRT and TensorFlow. Use TensorRT to accelerate TensorFlow graphs, even if the graph has TensorFlow operators that TensorRT does not support. TensorRT-supported subgraphs will be accelerated, resulting in a TensorFlow graph that can be executed as usual.

During TensorFlow optimization with TensorRT (TF-TRT), TensorRT performs some important transformations and optimizations on the neural network graph. First, eliminate the unused layers from the output to avoid unnecessary calculations. Next, the convolution layer, bias layer, and ReLU layer are fused together as much as possible to form a single layer. Another transformation is horizontal layer fusion or layer aggregation, and the steps required to partition the aggregation layers into their respective outputs. Horizontal layer fusion combines layers with the same source tensors and applying the same operations with similar parameters to improve performance

TensorRT provides a simple API that allows TensorFlow to use the 16-bit floating point and 8-bit integer optimizations in TensorRT. Tested against ResNet-50, TensorRT improves TensorFlow inference speed by eight times.

Let's take a look at the workflow and use some examples to get you started.

Optimizing TensorFlow Subgraphs

TensorFlow integration with TensorRT optimizes and executes compatible subgraphs. While using TensorFlow's extensive and flexible capabilities, TensorRT will parse the model and optimize possible subgraphs, leaving TensorFlow to execute the rest of the graph. TensorFlow programs require only a few lines of code to implement integration. Applying TensorRT optimizations to a TensorFlow model requires exporting the graph, and you may need to manually import some unsupported TensorFlow layers.

Let's walk through the workflow step by step. First, freeze the TensorFlow graph and request TensorRT to optimize TensorFlow's subgraphs. Next, TensorRT replaces each supported subgraph with an optimized node, and finally generates a frozen graph that can be used for inference in TensorFlow.

Figure 10.3 illustrates the workflow.

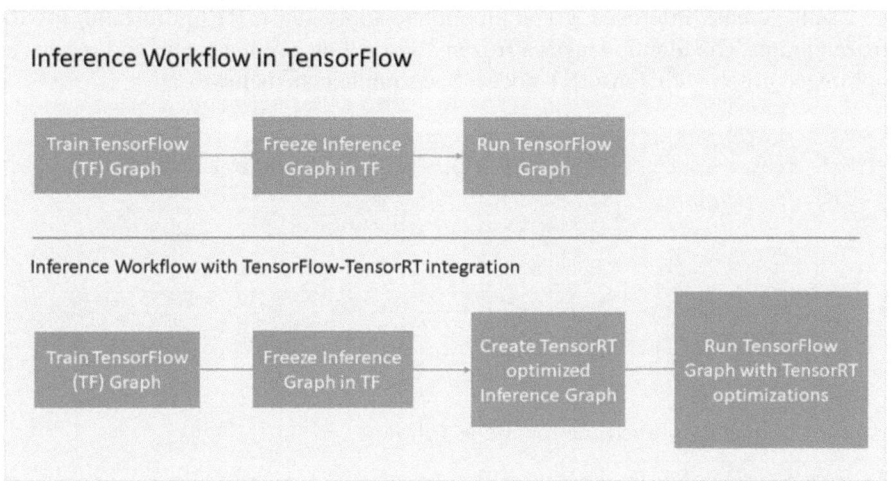

Fig. 10.3 Workflow diagram for integrating TensorFlow and TensorRT for inference

TensorFlow calls TensorRT to perform the optimized nodes and performs the rest of the graph itself. For example, let's say your graph has three segments, A, B, and C. Segment B is optimized by TensorRT and replaced by a single node. During inference, TensorFlow executes A, then calls TensorRT to execute B, and then TensorFlow executes C. From a user perspective, you will continue to use TensorFlow. Let's look at an example of applying this workflow.

Application Examples

Call the API in existing TensorFlow code to apply TensorRT optimizations to TensorFlow graphs. It can:

1. Specify the proportion of GPU memory used by TensorFlow. TensorRT can use the remaining memory.
2. Let TensorRT analyze the TensorFlow graph, apply optimizations and replace subgraphs with TensorRT nodes.

Use the per_process_gpu_memory_fraction parameter of the GPUOptions function to specify the share of GPU memory that TensorRT can consume. This parameter should be set when the TensorFlow-TensorRT process is first started. For example, set to 0.67, TensorFlow will allocate 67% of GPU memory, leaving the remaining 33% available to the TensorRT engine.

```
gpu_options = tf.GPUOptions ( per_process_gpu_memory_
fraction = number_between_0_and_1)
```

Use the create_inference_graph function to apply TensorRT optimizations to the frozen graph. The function takes a frozen TensorFlow graph as input and returns an optimized graph with TensorRT nodes. See sample code below:

```
trt_graph = trt.create_inference_graph (input_graph_def =
frozen_graph_def ,
                outputs= output_node_name ,
                max_batch_size = batch_size ,
                max_workspace_size_bytes = workspace_size ,
                precision_mode =precision)
```

The parameters of the function are as follows:

1. frozen_graph_def: frozen TensorFlow graph
2. output_node_name: Output node name string list, such as "resnet_v1_50/predictions/Reshape_1"
3. max_batch_size: Integer, enter the size of the batch, for example, 16
4. max_workspace_size_bytes: Integer, the maximum GPU memory that TensorRT can use
5. precision_mode: string, precision, allowed values are "FP32", "FP16," or "INT8"

The per_process_gpu_memory_fraction and max_workspace_size_bytes parameters should be used together for best overall performance. For example, for a 12 GB GPU, the per_process_gpu_memory_fraction parameter should be set to $(12 - 4)/12 = 0.67$, the max_workspace_size_bytes parameter should be set to 4,000,000,000, thus allocating approximately 4 GB of memory to the TensorRT engine.

Call Hardware Acceleration

Using half-precision (also known as 16-bit floating point) algorithms can reduce the memory usage of neural networks compared to 32-bit floating point or 64-bit floating point. This allows larger networks to be deployed with less time spent on transmission. NVIDIA Volta Tensor Core provides a $4 \times 4 \times 4$ matrix processing array that performs the operation $\mathbf{D = A * B + C,}$ where $\mathbf{A, B, C}$ and $\mathbf{D}$ are 4×4 matrices, as shown in Fig. 10.4. The inputs $\mathbf{A}$ and $\mathbf{B}$ of matrix multiplication are 16-bit floating-point matrices, while the accumulation matrices $\mathbf{C}$ and $\mathbf{D}$ can be 16-bit or 32-bit matrices.

When using half-precision algorithms, TensorRT automatically calls Tensor Cores in Volta GPU for inference. Taking NVIDIA V100 as an example, the peak performance of half-precision is about an order of magnitude (10 times) faster than double precision (64-bit floating point) and about 4 times faster than single precision (32-bit floating point). Since NVIDIA Jetson Xavier uses the same Volta architecture GPU as V100, the same results can be obtained on Jetson Xavier.

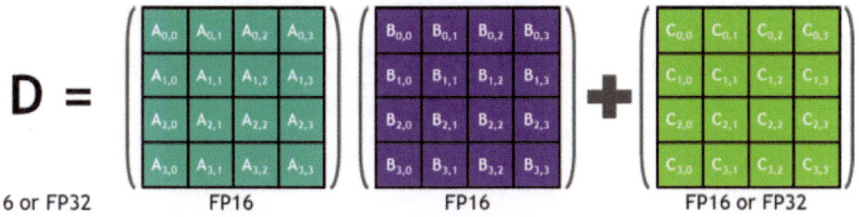

FP16 or FP32 FP16 FP16 FP16 or FP32

Fig. 10.4 Matrix processing operations on Tensor Core

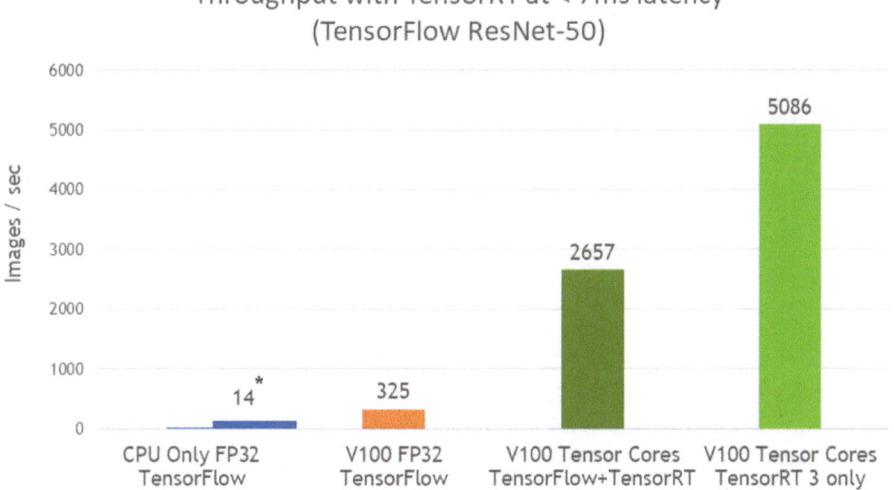

Fig. 10.5 Performance comparison of ResNet-50 inference before and after TensorRT optimization

Note: In the create_inference_graph function, set the precision_mode parameter to "FP16" to enable half-precision.

As shown in Fig. 10.5, the TensorFlow-TensorRT integration using NVIDIA Volta Tensor Core improves the performance of the ResNet-50 network by 8 times with a latency of less than 7 ms compared to TensorFlow.

8-Bit Integer Quantization

Performing inference with 8-bit integer precision can further increases computation speed and reduces bandwidth requirements. However, after the dynamic range is reduced, the accuracy of the weights and activation values of the neural network is lost. Table 10.1 illustrates this effect.

Table 10.1 Comparison of
dynamic range between
FP32, FP16, and INT8

	Dynamic range	Minimum positive value
FP32	$-3.4 \times 10^{38} \sim +3.4 \times 10^{38}$	1.4×10^{-45}
FP16	$65504 \sim +65504$	5.96×10^{-8}
INT8	$-128 \sim +127$	1

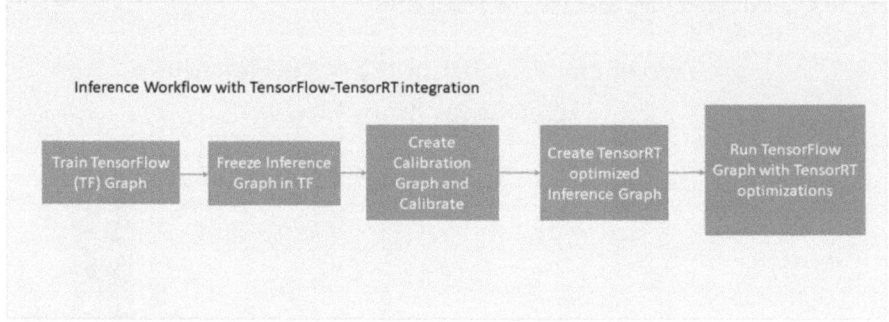

Fig. 10.6 Workflow for INT8 calibration before TensorRT optimization

To minimize accuracy loss, TensorRT provides the following functionality: train models in single precision (FP32) and half-precision (FP16), and then convert them into deployment models through INT8 quantization.

To convert the model to INT8 for deployment, the trained FP32 model needs to be calibrated before TensorRT's optimization, and the rest of the workflow remains unchanged. Figure 10.6 shows the workflow for adding the quantization step.

Quantization requires a calibration data set, and the following work is completed during quantization :

- Inference was performed using the original model (FP32 accuracy) on the calibration data set.
- Gather the required statistics.
- The optimal scale factor is obtained through the calibration algorithm.
- Quantize the weights of 32-bit floating point numbers into 8-bit integers.
- Create a "calibration table" and 8-bit integer execution engine.

In terms of program implementation:

First, set the " **INT8** " **calibration mode** using the precision_mode parameter of the create_inference_graph function. The function outputs a frozen TensorFlow graph.

```
trt_graph = trt.create_inference_graph(getNetwork(network_
file_name), outputs, max_batch_size=batch_size, max_workspace_
size_bytes=workspace_size, precision_mode="INT8")
```

Fig. 10.7 TensorRT integrated with Pytorch

Next, use the calibration data to calibrate the plot. TensorRT uses the distribution of node data to quantify the weight of nodes. It is important to use calibration data that reflects the distribution of the actual production data set. It is recommended to check the cumulative error during inference when using an INT8 calibrated model for the first time.

After executing the graph on the calibration data, apply TensorRT optimizations to the calibration graph using the function calib_graph_to_infer_graph. This function also replaces the TensorFlow subgraph with a TensorRT node optimized for INT8. The function outputs a frozen TensorFlow graph for use in inference.

```
trt_graph = trt.calib_graph_to_infer_graph ( calibGraph )
```

In this way, INT8 precision inference of the TensorFlow model is achieved using two commands.

Integrate with PyTorch

TensorRT can be used with (NVIDIA Corporation, 2020) PyTorch in the following ways, as shown in Fig. 10.7.

1. Convert the trained PyTorch model to ONNX.
2. Import the ONNX model into TensorRT.
3. Apply optimizations and build the engine.
4. Perform inference on the GPU.

This method consists of a specialized converter that converts PyTorch models into ONNX models. Currently, this converter has some limitations and cannot guarantee that all models can be converted successfully, especially complex models.

Another integration method can be used at this time, first using the TensorRT API to copy the network architecture, and then copying the weights from PyTorch (or other frameworks with weights compatible with NumPy types). Finally, perform the inference. This method is more complicated and requires developers to have a full understanding of the TensorRT API. Of course, the advantage is that it can give full play to the capabilities of TensorRT. When the original model has some complex operations, developers can find alternative implementations in the TensorRT API method to complete the model conversion.

10.2.2 Deployment to Embedded Systems

TensorRT can deploy trained networks to embedded systems (such as NVIDIA Drive and Jetson platforms). Deployment means taking the network and using it in software embedded in the device, such as object detection or image semantic segmentation. The deployment process consists of the following steps:

1. Export the trained network to a format like UFF or ONNX and import it into TensorRT.
2. Write a program that uses the TensorRT API to import, optimize and serialize the trained network into a solution file. For the sake of discussion, we will call this program make_plan.
3. (Optional) Perform INT8 calibration and export the calibration cache.
4. make_plan on the host system to validate the trained model before deploying to the target system.
5. Copy the trained network (and INT8 calibration cache) to the target system. Rebuild and rerun the make_plan program on the target system to generate the plan file. NOTE: The make_plan program must be run on the target system to properly optimize the TensorRT engine. However, if the INT8 calibration cache has been generated on the host, it can be reused on the target system, that is, there is no need to perform INT8 calibration on the target system.
6. Finally, applications for embedded systems use the TensorRT API to create engines from scenario files and perform inference.

10.3 Comparison of Two Model Optimization Techniques

For NVIDIA Jets on, both TensorFlow Lite and TensorRT provide model optimization capabilities, so what are the differences between them?

First of all, the advantage of TensorFlow Lite is that it provides training-time optimization tools, such as pruning, quantization training and clustering, but the disadvantage is that it is universal and the final execution code is not optimized for NVIDIA GPU (in fact, it does not support CUDA and CuDNN). The advantages and disadvantages of TensorRT are exactly the opposite. Its advantage is that it is specially optimized for hardware, so the compiled code execution efficiency is higher, but it does not provide the ability to optimize during training.

Secondly, from the convenience of use, TensorFlow Lite encapsulates the underlying implementation, making it easier to use, but it also loses some flexibility. TensorRT is just the opposite. It provides a more sophisticated underlying calling API, so developers can do more optimizations for hardware, but the problem is that they need to know more about the underlying technology.

Finally, in some actual tests, TensorRT's performance is slightly better than TensorFlow Lite.

It's also worth noting that installing TensorFlow Lite on the Jetson requires some skills.

AI applications with demanding operating environment requirements, TensorRT is recommended. If the requirements for inference performance and energy consumption are not so high, both are good choices. Of course, for devices such as smartphones, TensorFlow Lite is a better choice.

So, is it possible to combine the training optimization of TensorFlow Lite with the post-training optimization of TensorRT? The answer is yes. The former completes pruning and clustering work and outputs a model in TensorFlow format (such as a frozen graph), while the latter can use the model output by the former as input to perform quantization, compilation, model conversion, etc., and finally generates INT8 precision and N Model optimized for VIDIA GPU hardware!

References

NVIDIA Corporation. (2018). *NVIDIA TensorRT*. Retrieved from https://developer.NVIDIA.com/tensorrt

NVIDIA Corporation. (2020). *Speeding up deep learning inference using TensorRT*. Retrieved from https://developer.nvidia.com/blog/speeding-up-deep-learning-inference-using-tensorrt/

TensorFlow. (n.d.). *TensorFlow Model Optimization*. Retrieved from https://www.tensorflow.org/model_optimization

Chapter 11
Examples of Embedded Neural Network Application

Abstract This chapter takes the development of a drone-based sun umbrella as an example and comprehensively uses the embedded AI application development process and components introduced in the previous chapters to demonstrate how to develop an embedded AI application.

Keyword Drone-based sun umbrella

11.1 Application Scenarios

In this example, we will try to implement artificial intelligence applications in drones. Drone is a typical embedded software and hardware environment, which has the characteristics of small size, light weight, and strict energy consumption requirements. Implementing embedded artificial intelligence in drones undoubtedly presents significant challenges. At the same time, people have high expectations for the intelligence of drones. After all, as a smart device, it is not cheap, equipped with relatively powerful processors, cameras, and other equipment, and has the basis for realizing artificial intelligence. People hope that it will be like a flight assistant that can help people complete tasks that are difficult to achieve on their own. It can be used in personal consumption, industry, military, and other fields to achieve aerial photography, selfies, performances, power inspections, emergency rescue, pesticide spraying, aerial surveying and mapping, land and resources survey, pipeline inspections, maritime supervision, logistics and transportation, traffic control, meteorology monitoring, anti-terrorism and riot prevention, military reconnaissance, target attack, and other tasks.

With the rise of communication technologies such as 5G and satellite Internet, the communication speed and bandwidth of drones have been greatly improved. Some people envision using the edge computing capabilities of 5G or satellite Internet to empower drones and allow drones to operate autonomously. Human–machine realizes embedded artificial intelligence in cloud computing mode. The advantage of this method is that it can save the volume, weight, and power required

© Tsinghua University Press 2024

B. Li, *Embedded Artificial Intelligence*,

https://doi.org/10.1007/978-981-97-5038-2_11

for drones to carry AI modules. However, considering the characteristics of drones such as fast flight speed, wide range of activities, high response speed requirements, and all-weather operation, 5G or the delay and coverage capabilities of satellite Internet are difficult to meet its requirements. In terms of the obstacle avoidance capability necessary during flight, if we wait for the 5G base station or satellite to issue obstacle avoidance instructions to the drone, a collision may have already occurred. Therefore, embedded artificial intelligence in local mode on drones is a must.

When drones have AI computing capabilities, it opens a door to its application scenarios and can achieve more and richer functions in addition to the basic capabilities of shooting and transportation. Let's assume that we will implement a sun umbrella based on a drone. This umbrella can track people's movements in real time and block people from the sun at any time, thereby freeing people's hands and allowing you to happily go shopping and play in the hot summer.

Of course, it is not easy to achieve this. The drone is required to be able to track people's movements in real time and cannot let the sunshine on the owner due to slow response. There are many ways to track the human body with drones, such as using mobile phones or wearing bracelets on the human body. However, they all have their drawbacks. For example, when you pick up your mobile phone or wave your arms, if the drone dances with it, then you will be exposed to the sun from time to time. To achieve accurate tracking, it is best for the drone to have the ability to sense the contours and movement posture of the human body. After all, the purpose of shading is to prevent sunlight from hitting the head and upper body. And there is no more comprehensive way to track a human body than vision.

11.2 Hardware Selection

As can be seen from the above application scenarios, the drone umbrella has a camera that can photograph the human body (especially the upper body), and an embedded AI processor that can calculate the human body's outline and posture and predict the human body's movement direction. Of course, it also has the basic modules of a drone, such as flight control unit, motor, propeller, battery, etc. Finally, it comes with an umbrella. Here, the embedded AI processor is undoubtedly the core module.

Due to the size, weight, and energy consumption limitations of drones, we need an embedded AI processor that is small, light in weight, low in energy consumption, and high in performance. We review the test results in Part 2. Jetson Nano can achieve a high processing speed (FPS) when running various models, and the power consumption is only 5 W, which basically meets the performance requirements of AI computing. Of course, another important reason for choosing Jetson Nano is that it is a complete embedded AI computer. This board not only has a GPU but also an ARM CPU, memory, and camera interfaces, providing the essential components to run a complete AI application. At the same time, the integration level is high, and the overall size and weight are small. If you choose other AI accelerators, such as the ARM Ethos-N77, although it has a high performance of 5 TOPS/W and an energy consumption of only 0.8 W, the overall size, weight, and energy consumption will also increase. Another reason to choose Jetson Nano is that it has a good software ecosystem and can quickly develop AI applications.

11.3   Application Development

Next, we must select the AI model to realize the recognition and tracking of human body contours on Jetson Nano. For simplicity, we directly use the pre-trained human pose recognition model. If you need to train this model, it is a general technique that will not be explained in this book.

11.3.1   Download the Model

The pre-trained model files for human posture recognition available for download are given in Table 11.1. They are trained using PyTorch. The model files contain trained weights, and the definition of the model is provided through the source code.
   We choose the weight file resnet18_baseline_att_224x224_A. Its size is 81 MB. The processing performance on Jetson can reach 2 2FPS, which is enough to track a human body that does not move too fast.

11.3.2   Load the Pre-trained Model

Need to load a JSON file describing human posture, which describes human body parts (keypoint) and skeleton (skeleton). This JSON file is then used to create a human body topology tensor, which is an intermediate data structure that describes the connections between human body parts.

Table 11.1 Human posture recognition pre-training model

Model	(FPS) on Jetson Nano	Weights file size (MB)
resnet18_baseline_att_224x224_A	22	81
densenet121_baseline_att_256x256_B	12	84

JSON file is as follows:

```
{
    " supercategory ": "person",
    "id": 1,
    "name": "person",
    "keypoints": ["nose", "left_eye", "right_eye", "left_ear",
"right_ear", "left_shoulder", "right_shoulder", "left_elbow",
"right_elbow", "left_wrist", "right_wrist", "left_hip",
"right_hip", "left_knee", "right_knee", "left_ankle", "right_
ankle", "neck"],
    "skeleton": [[16, 14], [14, 12], [17, 15], [15, 13], [12,
13], [6, 8], [7, 9], [8, 10], [9, 11], [2, 3], [1, 2], [1, 3],
[2, 4], [3, 5], [4, 6], [5, 7] , [18, 1], [18, 6], [18, 7],
[18, 12], [18, 13]]
}
```

There are a total of 18 human body parts and 21 types of skeleton connections. Each skeleton connection is a connection between two human body parts. For example, [6, 8] represents a connection between left shoulder and left elbow, which is commonly known as the left arm.

```
import json
import trt_pose.coco
with open(' human_pose.json ', 'r') as f:
human_pose = json.load (f)
topology = trt_pose.coco.coco_category_to_topology (
human_pose )
```

Next load the model. Each model has at least two parameters, cmap_channels and paf_channels, corresponding to the number of Confidence Maps channels and Part Affinity Fields channels, respectively. The number of Part Affinity Fields channels is twice the number of skeleton connections, because each connection is a vector with corresponding channels in the x- and y-directions.

```
import trt_pose.models
num_parts = len ( human_pose [' keypoints '])
num_links = len ( human_pose ['skeleton'])
model = trt_pose.models.resnet18_baseline_att( num_parts , 2 *
num_links ).cuda().eval()
```

Next load the downloaded model weights file.

```
import torch
MODEL_WEIGHTS = 'resnet18_baseline_att_224x224_A.pth'
model.load_state_dict ( torch.load (MODEL_WEIGHTS))
```

11.3.3 Convert to TensorRT Format

The model loaded earlier is in PyTorch format. To run smoothly on Jetson, it needs to be converted to the optimized TensorRT format.

To use the python library torch2trt for TensorRT optimization, you also need to create some sample data. The dimensions of this data should match the dimensions of the trained network.

```
WIDTH = 224
HEIGHT = 224
data = torch.zeros ((1, 3, HEIGHT, WIDTH)).cuda()
```

Next, the model was optimized using torch2trt. We will start with FP16 mode with half precision. This is one of the advantages of Jetson Nano GPUs, other embedded AI accelerators can only run in the lower precision INT8 mode.

```
import torch2trt
model_trt = torch2trt.torch2trt(model, [data], fp16_mode=True,
max_workspace_size =1<<25)
```

The optimized model can be saved and simply loaded when executed again. Please note that TensorRT models are optimized for specific devices, and this optimized model can only be used on the Jetson platform.

```
OPTIMIZED_MODEL = ' model.trt '
torch.save ( model_trt.state_dict (), OPTIMIZED_MODEL)
```

The saved TensorRT model.

```
from torch2trt import TRTModule
model_trt = TRTModule ()
model_trt.load_state_dict ( torch.load (OPTIMIZED_MODEL))
```

Test the performance of model inference (FPS).

```
import time
t0 = time.time ()
torch.cuda.current_stream ().synchronize()
for i in range(50):
y = model_trt (data)
torch.cuda.current_stream ().synchronize()
t1 = time.time ()
print(50.0 / (t1 - t0))
```

If the performance reaches 20FPS, it will basically meet the needs of processing the human body in the video.

11.3.4   Inference

Next, the optimized model is called for inference: video is collected at a frame rate of 20FPS, preprocessed, inference is performed, the output is parsed, and the human body outline is drawn on the collected image.

```
import cv2
import torchvision.transforms as transforms
import PIL.Image
mean = torch.Tensor ([0.485, 0.456, 0.406]).cuda ()
std = torch.Tensor ([0.229, 0.224, 0.225]).cuda ()
device = torch.device (' cuda ')
# Define an image preprocessing function, the input of the
function is BGR8/HWC format

def preprocess(image):
    global device
    device = torch.device (' cuda ')
    image = cv2.cvtColor(image, cv2.COLOR_BGR2RGB)
    image = PIL.Image.fromarray (image)
    image = transforms.functional.to_tensor (image).to(device)
        image.sub _(mean[:, None, None]).div_(std[:, None,
None])
    return image[None, ...]
```

```
from trt_pose.draw_objects import DrawObjects
from trt_pose.parse_objects import ParseObjects
# Create a parsing object for parsing the output of the
neural network
parse_objects = ParseObjects (topology)
# Create a drawing object for drawing human posture
recognition results
draw_objects = DrawObjects (topology)
# Create a camera and capture images in real time at 20FPS
# from jetcam.usb_camera import USBCamera
from jetcam.csi_camera import CSICamera
from jetcam.utils import bgr8_to_jpeg
# camera = USBCamera (width=WIDTH, height=HEIGHT, capture_
fps =20)
camera = CSICamera (width=WIDTH, height=HEIGHT, capture_
fps =20)
camera.running = True
# Execute the neural network and draw human body outlines
image = camera.value
data = preprocess(image)
# Model output Part confidence maps and Part affinity fields
cmap , paf = model_trt (data)
cmap , paf = cmap.detach().cpu(), paf.detach().cpu ()
# Analyze and generate human body parts ( objects) and their
confidence levels ( peaks)
counts, objects, peaks = parse_objects ( cmap , paf )#,
cmap_threshold =0.15, link_threshold =0.15)
# Draw a human skeleton diagram based on human body parts (
objects) and human body topology ( topology)
draw_objects (image, counts, objects, peaks)
# Displays an image with human silhouettes superimposed
import matplotlib.pyplot as plt
image = cv2.cvtColor(image, cv2.COLOR_BGR2RGB)
plt.imshow (image[:, ::-1, :])
```

The output is shown in Fig. 11.1.

You can also print out the locations of various human body parts:

Fig. 11.1 Human posture recognition example

```
height = image.shape[0]
width = image.shape[1]
K = topology.shape[0]
count = int(counts[0])
K = topology.shape[0]
for i in range(count):
    obj = objects[0][i]
    C = obj.shape[0]
    for j in range(C):
        k = int(obj[j])
        if k >= 0:
            peak = peaks[0][j][k]
            x = round(float(peak[1]) * width)
            y = round(float(peak[0]) * height)
            print(human_pose['keypoints'][j], x, y)
```

Example output is as follows:

```
nose 114 60
left_eye 116 57
right_eye 113 57
left_ear 119 59
right_ear 111 59
left_shoulder 124 73
right_shoulder 109 75
left_elbow 132 74
right_elbow 106 94
left_wrist 133 67
right_wrist 105 111
left_hip 121 113
right_hip 111 114
left_knee 122 144
right_knee 112 144
left_ankle 121 174
right_ankle 113 174
neck 117 74
```

Based on the interrelationship between these parts and the time changes, it is not difficult to analyze the movements produced by the human body, such as normal walking and turning. In this way, according to the outline of the human body, letting the drone sun umbrella stay at a certain distance from the human head, and then detecting the changes in the outline of the human body's joints in real time, it is possible to predict in which direction the person will move forward, turn, go up or downhill, or even retreat. This allows the drone sun umbrella to track this movement and adjust the position of the drone umbrella at any time to achieve the following and sunshade effect on the human body. This model can also capture the postures of multiple people at the same time, combine it with face recognition, and then analyze the interaction between them, such as discovering that the owner has brought a family member/friend to go shopping together, or discovering that a curious stranger is ready to snatch the drone, etc., and ultimately achieve more interesting applications.

It should be noted that the above examples cannot yet create a practical drone sun umbrella. This is mainly due to battery power consumption. The battery life of the drone prevents this kind of umbrella from working for a long time. However, due to the rapid development of embedded artificial intelligence software and hardware technology, this example will not remain a mere science fiction idea. I believe that soon, this kind of drone umbrella can become a reality!

Chapter 12
Conclusion: Intelligence in Everything

Abstract This chapter is the summary of this book *"Embedded Artificial Intelligence."* In one sentence, embedded neural networks will empower everything with intelligence!

Keywords Embedded AI · Intelligence in everything

At the end of this book, let us review the origins and development of embedded artificial intelligence. After decades of exploration, people have finally found a key to unlock the door to artificial intelligence, and that is deep learning. By simulating the neural networks of the brain and the methods of learning new knowledge, people have found that it is possible to "replicate" the intelligence of the brain. Although we have not fully understood how all of this happens in a neural network that is deeply layered and highly complex, one thing we can be sure of is that intelligence is not magical. It is the product or external manifestation of the neural network of the brain, and the brain can be simulated by artificial neural networks. When we simulate it closely enough, miracles suddenly occur, and intelligence begins to break free from the constraints of the human brain, achieving realization on computers!

This gives us greater confidence. Since intelligence can be "transplanted," it can be transplanted onto large, high-level general-purpose computers, so it must also be possible to transplant it into small, ubiquitous embedded computing systems, thereby endowing various devices with intelligence. They can interact more intelligently with the external world, not just simply collecting data and performing programmatic control, but also analyzing, making decisions, and taking actions autonomously.

To achieve this dream, people began to attempt to shrink or tailor existing artificial intelligence from general-purpose computers and stuff it into the tiny brains of embedded computing systems. This seems less advanced and not as complex. Isn't it just throwing low end, less performance-intensive neural networks to single-chip systems on embedded devices to process? If it doesn't work, just let the embedded device only serve as eyes and ears and leave the rest to the cloud computing center.

© Tsinghua University Press 2024

B. Li, *Embedded Artificial Intelligence*,

https://doi.org/10.1007/978-981-97-5038-2_12

However, reality is not so simple. When we delve into the areas where embedded artificial intelligence is to be applied, we find that it is full of challenges. We need to execute ultra-high-energy-efficient, super-simplified algorithms and complete them in ultra-short times, and then place them in devices that are extremely small in size and weight. In short, we need to implement artificial intelligence in the most stream-lined way possible! This challenge is no less than letting general-purpose computers explore more complex artificial intelligence.

To overcome this challenge, people began to address the problem at various levels. In summary, to achieve embedded artificial intelligence, we need five components:

1. Embedded AI chips, which are AI accelerators with higher energy efficiency.
2. Lightweight AI algorithms with lower computational complexity and fewer parameters, but whose accuracy is no less than that of large-scale AI algorithms.
3. Model compression, which further prunes redundant parameters in lightweight AI algorithms and expresses them in a more concise manner.
4. Compilation optimization, which translates models into encoding that is more suitable for AI accelerator instructions.
5. Hierarchical cascade application frameworks that use the most suitable algorithms at the right time, thus reducing the cost and energy consumption of the entire system.

Combining these five components, embedded artificial intelligence begins to enter a rapid development trajectory. From the initial cloud computing mode to the local computing mode, of course, a device-cloud collaboration mode can be used as a transition. In the future, it may evolve into a cross-device joint learning model.

After achieving the goal of being small and powerful, embedded artificial intelligence hopes to go further and overcome a major flaw of traditional deep learning—the loss of learning ability after deployment. The reason why human intelligence is powerful is not only because it has the ability to induce from experience but also because it has the ability to quickly respond to unknown situations and learn new knowledge from them. If embedded devices have this lifelong learning ability, they will undoubtedly step into a broader future. After endowing everything with the ability of lifelong learning, they will truly "come alive" and become another form of life—artificial life.

Yes, artificial intelligence will no longer be confined to the virtual world, relying solely on processed and filtered information. It will break through all barriers and enter the real physical world, directly handling massive, high-speed, diverse, and entirely new streams of big data, truly understanding the real world, and even transforming the real world.

In short, embedded neural networks will empower everything with intelligence!